T0073804

GREENHOUSE PLANET

GREENHOUSE PLANET

How Rising CO_2 Changes Plants and Life as We Know It

LEWIS H. ZISKA

Columbia University Press
New York

Columbia University Press
Publishers Since 1893
New York   Chichester, West Sussex
cup.columbia.edu

Library of Congress Cataloging-in-Publication Data
Names: Ziska, Lewis H., author.
Title: Greenhouse Planet: How Rising CO_2 Changes Plants and Life as We Know It / Lewis H. Ziska.
Description: New York : Columbia University Press, [2022] | Includes bibliographical references and index. Identifiers: LCCN 2022004134 | ISBN 9780231206709 (hardback) | ISBN 9780231556613 (ebook)
Subjects: LCSH: Plants—Effect of atmospheric carbon dioxide on. | Atmospheric carbon dioxide—Environmental aspects. | Human-plant relationships. | Plants and civilization. | Climatic changes. | Science—Political aspects.
Classification: LCC QK753.C3 Z57 2022 | DDC 581.7—dc23/eng/20220331
LC record available at https://lccn.loc.gov/2022004134

Columbia University Press books are printed on permanent and durable acid-free paper.
Printed in the United States of America

Cover design: Julia Kushnirsky
Cover photo: J. K. York/Shutterstock

To those educators who took a chance on me,
my sincere and heartfelt thanks

CONTENTS

PREFACE

IN 1998 I found myself outstanding in my field.

Of soybeans, in Maryland.

"No, move a little bit to the front of the soybeans," the producer told me. I did my best to oblige.

"OK, now turn to me, but don't look at the camera directly." My head swiveled accordingly.

"Now what?" I asked.

"Read this," he said and inserted a sheet of talking points into my right hand. I glanced at the sheet and gave it back.

I looked over at the U.S. Department of Agriculture (USDA) National Program Leader, Dr. Herman Mayeux, standing a step or two behind the producer, and arched my eyebrows in a question. He nodded encouragement. As a postdoc hoping for a permanent USDA scientist position, I took any nod from someone who had my career interests at heart as a good sign. When I finished "not looking at the camera directly," there was some discussion between Herman and the producer. Herman stepped over.

"Can you talk about rice?" he asked, referring to my recent research at the International Rice Research Institute in the Philippines. Fair enough, I was—I am—passionate about rice.

I took a breath and began describing how more carbon dioxide (CO_2) in the atmosphere could stimulate rice growth.

When the filming was over, I asked the producer if I could also talk about how increasing CO_2 would make weeds in rice systems grow more. He smiled sympathetically and said, "Maybe another time."

Now cameras were being returned to large felt-lined boxes, doors were closing with a thunk, and engines were starting. "What's this all about?" I asked Herman.

"Something Sherwood Idso set up," he explained.

OK, I thought. Idso was a veteran USDA scientist from Arizona and had been doing CO_2 work for years. This was some sort of educational video. No worries.

Educational video? Not exactly. It was a documentary, or rather a sequel: *The Greening of Planet Earth Continues*. The original video, released in 1992, was *The Greening of Planet Earth*. In both videos, "newscasters" (real fake news!) interview scientists from around the world, marginalize the impact of global warming, and emphasize how great more atmospheric CO_2 will be for plants, from stimulating growth to increasing water use efficiency.

These videos were influential—instrumental in helping to create the myth that higher levels of CO_2 released via industry were wonderful, a boon to the environment. Thousands of copies of both videos were distributed to movers, shakers, and policy-makers by the coal industry.

And the message was simple: More CO_2 would be positive for plants, crops, the food supply, human health, and all of civilization.

That pro-CO_2 message from 1998 has not gone away. The "CO_2 is plant food" meme continues on, having provided an essential talking point for climate change deniers for the last three decades. Of course, the clumsy VCR videos have been replaced by YouTube, but the message remains the same.[1] And rest assured that if any public testimony regarding climate change is needed, the "CO_2 is plant food" message will be repeated.

Who are the individuals and groups promoting this idea? We can begin with Sherwood Idso, the former USDA scientist who narrated both videos under the auspices of the Greening Earth Society. "Former" because with money from the Western Fuels Association—that is, the coal industry—he started his own institute, the Center for the Study of Carbon Dioxide and Global Change. While the Greening Earth Society is now defunct, the Center for the Study of Carbon Dioxide and Global Change is still operational. Sherwood's initial efforts are continued by his sons, Craig and Keith, who run the website http://co2science.org, which takes all scientific papers regarding CO_2 and plant biology and interprets them according to their own ideology. Simply put, they remove any conclusions that don't agree with the premise that an increase in CO_2 is wonderful and beneficial.

The Idsos are only the tip of a climate-denying iceberg. The CO_2 meme is ubiquitous in conservative politics. It is an involuntary mental tick in deniers' minds, as certain as a sharp tap on the patellar tendon causing the knee-jerk response. Should anyone utter, "Climate change is man-made," the response will issue from a conservative's tongue faster than any pulled string, "But CO_2 is plant food!"

And this is the conservative chorus: from Michele Bachmann, diminutive religious holy roller and ex-presidential candidate (2012), insisting that CO_2 is natural and that reducing it would hurt our standard of living, to Lord Christopher Monckton, the pop-eyed peer consultant and skeptic gadfly who, while having no background in science, has somehow managed to write "peer-reviewed" opinion pieces that insist that CO_2 is beneficial and that any policy to reduce its concentration will be destructive to society.

As for scientists? Well, two of the most prominent are Patrick Michaels, a former University of Virginia biologist who consistently disputes the science behind climate change, insisting that the more CO_2 we put in the atmosphere, the greener the earth will get, and William Happer, the octogenarian physicist from Princeton who has argued that the "demonization of CO_2 . . . differs little from the Nazi persecution of the Jews, the Soviet extermination of class enemies or ISIL['s] slaughter of infidels." (Really.)[2]

You could be forgiven for thinking that CO_2 is a holy cause among the conservative intelligentsia, that to disparage the role that CO_2 plays in plant biology is both un-American and the scientific equivalent of dismissing gravity.

I confess that when I received my copy of *The Greening of Planet Earth Continues* and popped it into the VCR to watch,

I didn't feel un-American or scientifically void. I felt like a dope, a tool of the first order. I was not the only one; I learned many years later that other scientists interviewed for the "documentary" also felt that their words had been taken out of context in an effort to further a political cause.[3]

In the pages that follow, I will try to bring a scientific perspective to the chorus of "CO_2 is plant food." I want to do so in part because all these conservatives are right: CO_2 *is* plant food.

How do I know? Because I am a plant biologist. I received my PhD in plant physiology from the University of California, Davis, in 1988 and have been looking at the impact of rising CO_2 on plants since 1989, studying Spartina along the Chesapeake wetlands as a Smithsonian fellow. In our "post-truth" and "alternative facts" world, expertise and experience are what I have to offer. But as any scientist would say, please don't rely on my word alone. Go to Google Scholar and type in "elevated CO_2" and "plant biology." You will find hundreds of peer-reviewed studies demonstrating that more CO_2 can stimulate plant photosynthesis and growth.

So is the phrase "CO_2 is plant food" the silver lining of climate change? A future in which plants are greener and the world transforms into a second Garden of Eden?

Without giving away the ending, the answer is an unequivocal no. So what does the increase in CO_2 *really* mean for plant biology? To answer that question, I want to explore the idea that "CO_2 is plant food" more deeply. To reveal what is behind the botanical curtain and to celebrate and explore how human existence is made possible only by plants, how plants are basic and beautiful, of benefit and bonus, but also of harm and hurt.

And I want to provide context. Sure, you've heard about CO_2 and climate change, of drowning polar bears, rising sea levels, of storms on steroids. All worrisome, of understandable concern, these are dramatic outcomes with Old Testament consequences. But we are so focused on this narrative that we have developed a blind spot when it comes to the simple yet powerful meme "CO_2 is plant food."

While the statement is true, the consequences of increasing levels of CO_2 are only now coming into view. And those consequences are potentially enormous, of a global impact rivaling anything you may have heard about climate change itself. Why? *Because the recent and projected increase in CO_2 will, by itself, directly and unequivocally affect all biology, no matter what climate change does.* The fact that CO_2 will affect life directly is not derived from new science but from very old knowledge: the recognition that plants supply food, energy, and carbon to all living things.

"CO_2 is plant food" has consequences, and those consequences are incredibly complex, potentially much more formidable than climate change per se, and will affect—directly, fundamentally, and irrevocably—all life as we know it.

PART I

A GREEN BLINDNESS

Chapter One

PLANTS ARE IMPORTANT

THE PART ABOUT FOOD

YOU CAN be forgiven if, while cruising down an interstate at 70 mph, your response to a herd of animals moving against a green background is "Look at the cows, Mom!" or sticking your head out the window and in a low voice calling, "MOOO." Prior to handheld, plastic, electronic miracles, yelling "MOO" at cows was my family's favorite means of making time pass on long car trips.

Did we comment on the green background? Not so much. No one in the car exclaimed, "Ooh, look at the birch . . . and that chestnut! That's exceptional! And check out the basswood! Isn't that bluegrass and ryegrass the cows are eating?"

Understandably, no. We are programmed to respond to movement because moving animals must be identified, almost at once, as friend or foe. The grass and trees? Meh.

We have keen abilities, hardwired via evolution, to identify the immediate, to evaluate threats and circumstances. But we also suffer from a green blindness, an inability to place trees, shrubs, and flowers in context, to identify them, to evaluate them. That green blindness is not universal; we can identify the small subset of available plants that affect our well-being. At the supermarket we can differentiate carrots from cauliflower, and our hunter-gatherer ancestors could do much more than we can in that department. (I doubt if many early hominids perpetually brushed up against poison ivy.)

But while we can seek the familiar in plants, few of us recognize their importance, their essential omnipresence. Rather, humans, our chests puffed out, hands on hips, feet astride manicured lawns, claim to be at the apex of evolution. After all, *we* cover the entire surface of the earth; *we* reside in every ecosystem. Our DNA is highly evolved, much more sophisticated—more than three billion base pairs!

Plants? Well, let's see. Huh. They also cover the entirety of the earth's surface, except *they* constitute 99 percent of all living biomass—not animals, and certainly not humans.[1] *Plants* are the more complete, the more complex organism. DNA? How about *Paris japonica*, a rare Japanese flower? It has 149 billion base pairs—fifty times more DNA than humans.[2]

There's more. In the simplest, most authentic of terms, if plants did not exist, neither would animals. Plants provide oxygen and move millions of tons of water from soil to air, forming cloud cover and indirectly creating their own weather patterns. Plants affect soil, their roots adding all the necessary ingredients—carbon, nitrogen, and phosphorus—to generate muddy microcosms: whole systems of millions of bacteria, fungi, and nematodes. Aboveground, plants are the basis of all ecosystems, from prairie to rainforest, from tundra to desert. No animal species—from aardvarks to zebras—can survive without plants.

And human history is enmeshed with plant biology in the most intimate and most telling of ways. We are ruled, dominated, and controlled by plants. Civilization would not exist without the efforts of proto-agriculturalists to collect certain cereals in river valleys around the world. Valleys whose river waters ebbed and flowed, depositing the silt and nutrients necessary to support primitive strains of wheat, barley, and rice. Valleys that, in turn, begat inverted human hierarchies in which priests and merchants, philosophers and traders, kings and bishops vied for standing at the top of the social pyramid. A pyramid made possible by those whose sweat and blood allowed the elites to avoid toiling for sustenance.

And our history begins. A hundred farmers, collectivized plant managers, even if poorly nourished by a single cereal species, could still dominate small tribes of hunter-gatherers—just as the latter group, relying on a more diverse set of plants, were taller and nutritionally superior.[3] As managing plants made it easier for one to feed many, surplus populations came to reflect the desires of leaders, whose priorities shifted to conquering new lands and finding new resources.

Boom: a good year, lots of plants, lots of food, more procreation, more people, more influence. As other groups were conquered, the influence of successful groups spread.

Bust: a bad year for plants, crops fail, people starve. More bad years? Desperation. Diasporas. Dissolvement. Histories can't be written by ghosts.

Our antiquity as a species is illustrated—illuminated—delineated by contractions and expansions of plant growth, by the civilizations that arose from the valleys where such growth occurred, from Mesopotamia to Egypt, from the Indus Valley to the Yangtze River basin.[4]

As children, we learn of the glories of the ancient Egyptians: their pyramids, their writing, their language, their culture.[5] Such glories reflect their environment, typified by the majestic Nile, the world's longest river, overflowing its banks each year in September, creating a floodplain a dozen miles wide and depositing new fertile silt into natural basins so that fertilizer was not necessary. And doing so in a land of eternal sunshine. Providing the means to cultivate the plants necessary for civilization: emmer (wheat), chickpeas, lentils, lettuce, onions, garlic, barley, papyrus, and flax. By around 5000 BCE, as the population increased, the need arose to domesticate animals like cattle, who also fed on the abundance of plant life. And from this consistent source of plant life emerged many of the awe-inspiring elements of ancient Egyptian culture: writings, medicines, mathematics, architecture (temples and obelisks), shipbuilding, makeup, and toothpaste, the hallmarks of civilization.

And armies. Young men fed by sustained plant growth (and furnished with the products of the discovery of new metals) could march and overcome their neighbors. So isolated tribes became towns. Towns, cities, then dynasties.

Yet what plants provide, they can take away. The Old Kingdom of Egypt began with the Third Dynasty in 2686 BCE and ended with the Eighth Dynasty more than five hundred years later. The glories of this kingdom, including the great pyramids of Giza, remain a global wonder.

But this kingdom depended on plants. And ancient writings indicate that Egypt was experiencing a lack of Nile flooding toward the end of the Old Kingdom period. No floods, no deposits of fresh nutrients, and no return of the plants necessary to sustain the Egyptians' empire. Upheaval occurred, and a new kingdom, the Middle Kingdom, was born.

The reasons for the iterations of ancient Egypt and its final collapse and domination by the Roman Empire are many, but the lack of plants, their contraction following the lack of flooding by the Nile, was a contributor.

The Egyptian scholar Ipuwer provides the following horrific description of the period of the Middle Kingdom when flooding had once again stopped:

> Lo, the desert claims the land. Towns are ravaged, Upper Egypt became a wasteland. Lo, everyone's hair [has fallen out]. Lo, great and small say, "I wish I were dead." Lo, children of nobles are dashed against walls. Infants are put on high ground. Food is lacking. Wearers of the linen are beaten with [sticks]. Ladies suffer like maidservants. Lo, those who were entombed are cast on high grounds. Men stir up strife unopposed. Groaning is throughout the land, mingled with lament. People are diminished.[6]

That plants, or the lack thereof, have toppled empires is not news. From the droughts that killed the plants needed to sustain the Ancestral Puebloan culture in the southwestern United

States in the twelfth century, to the bread riots preceding the French Revolution in the 1770s, to the diaspora of the Irish to America following the loss of potato crops to a plant fungus (*Phytophthora infestans*) in the 1840s, we can see that when plant crops go missing, human lives suffer.[7]

How much suffering is caused? That is the subject of some educated guesswork, as records of the number of deaths from famine were not recorded in ancient Egypt or any of the civilizations that emerged following the adoption of plant cultivation. However, in the book *Food or War*, the science writer Julian Cribb tries to provide an answer, beginning with eighteenth-century India and continuing through recent famines in Africa.[8] The estimated number is staggering—more than two hundred million deaths—with the greatest famine event being from 1958 to 1962, when about forty-five million people died during China's Great Awakening.

That event is a blunt reminder that plant insufficiencies can be caused by the environment: dwindling rainfall, rivers no longer flooding, extreme heat, and drought make grim partners for plant growth. But insufficiencies can also result from humans: their incompetence, their politics, their fear, their hatred, their prejudices. Such characteristics can, and often do, exacerbate plant shortages. When humans wrote of the Four Horsemen of the Apocalypse, there was a reason famine and war were included.

But hey, this is the twenty-first century. C'mon. Sure, we still need food, and plants provide that, but that kind of historic dysfunction is a thing of the past, right? I mean, my fridge is full of food, and I've got three supermarkets within walking distance. That old hand-to-mouth existence is over.

Let us return to Egypt, now a country of ninety-nine million, a population no longer sustainable by its own plant-growing efforts. When once Roman armies depended on wheat imports from Egypt, Egypt is now among the world's largest importers of wheat. Arab countries, home to many of the world's early dynasties, now rely on bread as a cheap source of food, available to even the poorest families.

In the summer of 2010, temperatures in Europe and western Russia spiked. Soon, heat and drought, with unprecedented fires and a string of days with temperatures above 100 degrees Fahrenheit, became the norm. More than eleven thousand deaths in Moscow were recorded.[9] In southern Russia and Ukraine, primary wheat-growing regions, with much of the wheat slated for export, production collapsed. Soon, with concern for their populations very much on the minds of authorities in these countries, all wheat exports were banned. That summer the price of wheat soared by 90 percent, exacerbated by unprecedented flooding on wheat grown in the southern plains of Canada.[10] By early 2011, Algeria was rocked by widespread protests. Tunisia's government was overthrown. In Egypt, President Hosni Mubarak was arrested. With the lack of one particular plant, wheat, the Arab Spring was born.[11]

When plant stress occurs, human societies that are already fractured are most at risk. Among Middle Eastern societies there is Syria, with its long history of regional and tribal cultures. Plants cannot be grown in a drought, so farmers in eastern Syria—1.5 million of them—moved to cities, including Aleppo and Damascus. Plants and water were scarce, and conflicts multiplied. Political leaders ignored the plight of these farmers and denied them resources needed to live. Civil war erupted. About half the Syrian population, roughly five million people, left the country, with consequent political fallout, increasing xenophobia, and worries of unrest gripping Europe and the rest of the world. Because there was not enough wheat. Not enough food. And the relationship between plants and humans continues unabated. Of the 68.5 million refugees in the world in 2018, most were fleeing because the plants needed to sustain them were in short supply.[12]

Plants are food. Anything that affects how they grow—and what we do in response—has had consequences since civilization began. Those consequences will continue until civilization ends.

Chapter Two

PLANTS ARE IMPORTANT

THE PART ABOUT DRUGS

CHANCES ARE that not far from where you lay your head at night, there is a drugstore, an apothecary, a gallipot where medicines are available. Headache? Muscle pull? Back pain? No worries. Pop in, pop out. Pop pill. Problem solved, or at least, eased.

We fool ourselves into thinking that our experiences are common to people all over the world. But many people, most of the world in fact, have never seen a modern drugstore. Ironically, in many urban areas, people also lack access. It turns out that if there are food deserts where fresh produce isn't always available, there are also pharmaceutical deserts such as in Chicago, where a million people don't have access to a corner drugstore.[1]

But if medicine does not come prepackaged and sanitized, it still exists. Drugs are not produced in a magic room in the back of the Rite Aid or Walgreens; rather, they arise from nature, and as nature consists primarily of plants, plants remain a primal source of new chemistry.

And this is by no means surprising. As humans we are motion oriented, and plants seem to remain static, an unchanging, green baize. But if we tuned in to our chemical senses, if they dominated our brains, plants would represent a moving kaleidoscope of riotous colors. For the simple reason that plants, with few exceptions (e.g., the Venus flytrap and the dogwood bunchberry), are incapable of rapid movement. So how do they escape predators? How do they communicate? Heck, how do they have sex?

And the equally simple answer? Through chemistry. From the roots to the shoots to the flowers, plant chemical dynamics would make the CEO of DuPont ("better living through chemistry") blush.

A simple but self-motivated example. Caterpillars arrive and start eating your leaves. Not good. Can't pull up your roots and run away. What to do?

Easy. First you send out a scent signal from your leaves for help. Now parasitic wasps arrive, laying their eggs in the caterpillars and killing them. But wait, won't that chemical signal also keep away pollinators? No worries. You make more flowers to produce a different attractive smell. Problem solved. As described by the Swiss researcher Florian Schiestl of the University of Zurich, plants can create a bouquet of scents to attract and repel the insect community in a very targeted manner.[2]

A bouquet is a wonderful aspect of plant chemistry, a mental stimulant. But it is only the merest whiff of a plant's chemical potential. Internally, leaves, stems, roots, flowers, and seed tissue can produce a wide and extraordinary arrangement of some of the most complex, most intricate means to beguile and dominate human physiology. Here are a few examples. Caffeine, a mild stimulant consumed by more than 250 million Americans *every day*, is the world's most consumed psychoactive substance.[3] Another widely used stimulant is nicotine, consumed by more than one billion active smokers in the world.[4] Then there's aspirin, 50 to 120 billion pills of which are consumed globally each year, derived from salicylic acid found in jasmine, beans, peas, clover, and certain grasses and trees. Ethanol is derived from a wide array of plant sources, from corn to sugarcane, from sweet potatoes to barley—and when it isn't being used to supplement gasoline, it's being downed at the local bar on a Friday night in another multitude of forms, from ale to vodka. So many plant-derived chemicals modify our being: artemisinin, atropine, cincristine, codeine, cryptolepine, digitalis, ephedrine, etoposide, irinotecan, L-DOPA, lapachol, paclitaxel, quinine, scopolamine, thebaine, topotecan, vinblastine. We use them for first aid, for recreation, for celebration, to harm, to help, to ease, to poison, to prevent poisoning, to slow disease, to avoid disease altogether.

Want a cuppa? Or how about a pill? "The pill" is the most commonly used oral contraceptive, one that provides women with control over their reproductive choices.

It is 1943 when an eccentric American professor from Penn State arrives at a small company, Laboratorios Hormona, in Mexico City. The company specializes in purifying hormones from human organs, a difficult and expensive process. The professor, Russell Marker, is both a botanist and chemist and is interested in a class of plant products called sapogenins. His wanderings have led him to Mexico where he has harvested a local yam species, *Dioscorea mexicana*, which locals refer to as *cabeza de negro*. Back at Penn State, Marker perfects a means to make progesterone from the diosgenin in *cabeza de negro* and heads back to Laboratorios Hormona (after failing to get any U.S. pharmaceutical company interest in his product) with almost ten pounds of progesterone, worth about $10 million (in today's prices).

Voila! A new pharmaceutical company, Syntex, is born, but arguments precipitate Marker's leaving the firm. A few years later, Carl Djerassi, the chemist known as the "father of the pill," is hired. Syntex rapidly becomes the leading supplier of sex hormones, including progesterone, testosterone, and various estrogens, all made from diosgenin.[5]

The pill is a modern use of plant chemistry, and progesterone et al. are now synthesized within the laboratory; few recognize its plant origins. But other plant sources have been used historically in human attempts to control reproduction: from the seeds of Queen Anne's lace (wild carrot), the use of which Hippocrates described more than two millennia ago; to pennyroyal, which the ancient Greeks used in teas to induce menstruation and abortion; to blue cohosh (papoose root), used for birth control by Native Americans and still used by some midwives today in the last month of pregnancy to tone the uterus prior to labor.[6]

But after visiting a drugstore and viewing the clean shelves, rows of perfectly square boxes, fluorescent lighting, and cardboard cutouts, one can be forgiven for missing any association with nature, for wondering if plants have *anything* to do with human medicine. Perhaps this is the appeal of supplement stores, where images of plants (e.g., Saint-John's-wort, taken for depression) at least grace the outside of the boxes.

OK, modern medicine may use plant chemistry as a template, but surely modern medicine doesn't *still* rely on harvesting plants in the field?

Pain. George Orwell wrote in *Nineteen Eighty-Four* that in the face of pain, there are no heroes. Chronic pain is unforgiving, incessant, jagged, sharp. And we seek relief. Chemical relief. And plants, once again, fill the void. Specifically, poppy (*Papaver somniferum*), one of about 120 species of *Papaver*, only two of which, *P. somniferum* and *P. setigerum*, produce a unique class of chemicals called opioids.

And they *are* farmed—extensively—by large pharmaceutical companies like GlaxoSmithKline. Poppy is sown and grown in fields from India to Turkey to Australia (Tasmania). Morphine, codeine, and papaverine are derived unchanged from the plant as has been done for millennia. (The ancient Sumerians, tending plants along the Tigris and Euphrates rivers, referred to the poppy as "the plant of joy"—for good reason.)

The role of morphine in easing pain is undisputed. Vital. Critical. Its use in cancer wards and its administration on battlefields are indispensable. (When the Allies cut off Nazi Germany from morphine, Nazi scientist derived the first synthetic version of morphine, known today as methadone.)[7]

But as plants are the apex of chemical production, they do not distinguish between "good" and "bad" when it comes to human society. If they reward, they also punish. Plants insist on some of the most toxic substances as a necessary means to keep animals, including humans, at bay.

How toxic? Well, ricin is an analogue of cyanide, produced by castor bean plants. In its purified form, a few grains can kill you. Just ask Georgi Markov, the Bulgarian dissident killed in September 1978 via ricin, imaginatively inserted via umbrella into the back of his right leg.[8] Then there's the rosary pea, a shiny seed used to make beads and jewelry—nontoxic if unbroken but if scratched or chewed, it releases a chemical similar to ricin called abrin that can be equally lethal. A mere 3 micrograms (0.00011 ounces) can kill a person.[9]

Other plant poisons are historically more familiar such as hemlock. Sentenced to death by local authorities for introducing strange gods and corrupting Athenian youth, Socrates consumed a tea made of hemlock. Hemlock contains coniine, a poison similar in structure to nicotine but one that disrupts the central nervous system.

And the incredible diversity of plant chemistry, including that of the venomous plants, has been brought to the fore through human intervention in many ways. Perhaps one of the most fascinating examples relates to the original use of plants to relieve pain. Salicylic acid has long been used as a pain reliever around the world, from the ancient Sumerians to Native Americans.

In the late nineteenth century, a chemist named Felix Hoffmann working for the German pharmaceutical company Bayer knows about salicylic acid but also its side effects, including nausea. He decides to acetylate it using a pH modifier (e.g., vinegar), simply removing one hydrogen molecule and replacing it with an acetyl group.[10] And acetylsalicylic acid, or aspirin, is born. How important is aspirin? Aspirin is considered by global authorities such as the World Health Organization to be a vital part of any health system—an essential drug.[11] "Aspirin" is still Bayer's registered trademark in more than eighty countries. It has and will continue to be a money-spinner for Bayer for a long time to come.

Humans can amplify plant chemistry to a new high, or a new low. And a low can cause thousands of people to suffer and die. Plant addiction, from nicotine to cocaine to heroin, is prevalent, but no plant addiction surpasses that of ethanol. Ethanol is derived from plants and comes in many forms: Vodka is derived from potatoes, beer from barley, whiskey from grains, and wine from grapes. An estimated eighty-eight thousand people die from alcohol-related causes every year, making alcohol the third-leading preventable cause of death in the United States. (The first? Tobacco.)[12]

And so it goes. Our current "fashion plant" is hemp. Hemp, aka cannabis, and its three major species—sativa, indica, and ruderalis—have intertwined themselves in all aspects of human existence for millennia, and why not? Hemp is used to make rope, paper, canvas, protein powder, oil-based paint, solvents, insulation, soap, shampoo, and more. Heck, George Washington was a hemp grower.

The latest use for cannabis is cannabidiol, or CBD. There have been reports that it may help treat pain, insomnia, and anxiety; it may even reduce acne and help treat heart disease.[13] Whether CBD in fact does any, or all, of that is still open for debate, for additional testing, but, for the moment, there is no shortage of stores and stories touting such therapies. Another side of the hemp coin is tetrahydrocannabinol, or THC, a psychoactive substance that can benefit people with cancer and HIV by increasing appetite. It does not kill you, but like alcohol and nicotine, it can impair judgment and can lead to death.

Ultimately, plants are chemical Merlins, capable of deriving the most complex, the most sophisticated of substances: those that can cure, heal, relieve pain, and even regulate the human reproductive system. Yet plants can also produce compounds that can devastate the human psyche, producing cravings and desires that can *literally* drive us insane. The result is death and suffering beyond imagining.

While leaving us begging for more.

Chapter Three

PLANTS ARE IMPORTANT

THE PART ABOUT RELIGION

RELIGION, THE need to instill our instincts with a higher power, a power that controls the twists and turns, fates and consequences of daily actions—and reactions—is a basic aspect of human existence.

Religion varies by geography, by commerce, by ethics, by education. It is a sociocultural set of behaviors and practices, with forms that are different in all the ways that humanity differs. Yet it has one functional commonality: plants.

Plants are the source, the font of what nature provides, from food to medicine, so it is not surprising that they can transcend daily needs and become celestial, sacred. Plants occupy two essential aspects in global religion: the first is as a passageway, a conduit to spirituality; the second is one of symbolism, of rites and meaning, of impact.

In touching the divine, the "substance of things not seen," a deeper universal meaning, plants have been used to induce altered states of consciousness for thousands of years. Used as a means to generate the divine within oneself, a direct link to the universe. How these plants are used provides cultural insight, reflects ethical values, honors what a society holds dear. It is also fair to say that their use is carefully controlled. Precise. Because there is a thin chemical line between vision and death.

Psychoactive plants used within rituals include some of the familiar ones (poppy, cannabis) but also a distinct group of plant species. Henbane, *Hyoscyamus niger*, produces a number of psychotropic substances, including scopolamine, an analogue of ecstasy, which was used by priestesses of Apollo during Hellenic times as a means to offer oracles or prophecies.[1] *Silene undulata*, native to the eastern cape of South Africa, is regarded by the Bantu as a sacred plant whose roots can be used to induce vivid and prophetic dreams.[2] Salvia, specifically *Salvia divinorum*, whose leaves produce opioid-like compounds, is native to Mexico and was used by Mazatec shamans to enable visions and higher levels of consciousness as part of spiritual healing.[3]

It is easy enough to dismiss past use as "uncivilized." But past can become prologue. Santo Daime is a synthesis of precepts and beliefs originating in the Amazonia state of Acre in the 1930s in far western Brazil. Here the laity undergo a "cleansing" in which ayahuasca (aka Daime), a psychoactive liquid derived from the *Banisteriopsis caapi* vine and *Psychotria viridis* shrub, is given to heal body and soul.[4] Today, ayahuasca retreat centers throughout Brazil are a popular attraction, providing a high-speed highway to enlightenment.

Other psychoactive plant substances, such as peyote (*Lophophora williamsii*), are also popular. Established in the nineteenth century, the Native American church is a unique blend of Christianity and Indigenous ritual. Practitioners consider peyote and its effects to be a sacrament. Once popular in the counterculture heyday of the 1960s, peyote is regaining its allure (though perhaps it never went away) and has become legal in cities as diverse as Oakland, California, and Denver, Colorado. Indeed, there is even consideration, all sacrament status aside, to use peyote as a prosaic cure for the mental or emotional distress associated with alcohol addiction.[5]

For some in the Western world, it can be tempting to dismiss these plants as too exotic, too foreign, and anathema to the world's main religions. But this would be wrong. In Christianity, the use of myrrh and frankincense is well known and widespread. Frankincense is manufactured from the Boswellia tree found in dry mountainous regions from Africa to India. Myrrh is sap, or resin, collected from cuts in the bark of trees of the genus *Commiphora*. These plants were not used to induce visions; rather, the smoke from the burning incense is considered a means by which prayers of the faithful are conveyed to Heaven.[6] Plants in Christianity act as intermediaries between Heaven and Earth.

When not active in ritual, plants continue to play a prominent role as potent religious symbols. Mistletoe, eternally green and flowering at the winter solstice, is used in Druidic rituals to symbolize immortality on the shortest day of the year.[7] For Hindus, the lotus flower has deep spiritual meaning with its origins associated with the god Vishnu.[8]. For Hindus this beautiful flower, glorious and iridescent, white in purity, represents life, fertility, perfection. Yet it rises from the mud and murk, symbolic of one's daily life. Hindus believe that within each person the spirit of the sacred lotus resides. Buddhists see the lotus flower as indicative of life's progress, one's spiritual journey. A closed flower bud represents a time before enlightenment, whereas a fully open flower symbolizes explanation and education, progress toward Nirvana.[9]

If flowers are prominent in religion, so are trees. Buddhism acknowledges that the Buddha achieved enlightenment while sitting under a Bodhi tree (*Ficus religiosa*). Japan has groves associated with Shinto shrines, and the Druids also had tree groves, usually associated with their goddess Nemetona. In Norse lore, the Yggdrasil tree, an enormous ash, connects the roots of the dark underworld to the starlit heavens. In Judaism, four tree species are used (citron fruit, date palm, myrtle, and willow) to celebrate the holiday of Sukkot (the Feast of Tabernacles) as a mitzvah (commandment) to mark the end of the harvest (see Leviticus 23:40). And the tree of immortality, or tree of life, motif is featured prominently in the Quran.[10]

Christianity, of course, is rife with tree imagery. Perhaps the original symbol is the tree of knowledge, imparting good and evil—first to Eve and then Adam. The yew tree is another powerful symbol, representing resurrection (reflecting the nature of the yew, which can grow a new trunk within the hollow husk of an old tree). In days of Christian yore, there was a ritual of placing yew shoots within the coffins of the deceased. Indeed, it was customary to establish churches where yews were grown.[11]

Such plant-based rites also extend beyond trees. In Greek Orthodox Christianity, basil (derived from the Greek for "kingly" or "royal") is associated with worship of the cross, especially during Lent when the priest uses basil to purify the holy water and then uses the leaves to sprinkle the congregation. The cross, suitably decorated in basil, is then processed around the church, and small bunches of basil are passed out to be taken home, put in water until roots develop, and then planted as a blessing.[12] And palm branches, a symbol of goodness and victory, were often depicted on coins and buildings (e.g., Solomon's Temple) during Roman times as a means to welcome Jesus to Jerusalem.

Perhaps the simplest but most meaningful way of using plants in a religious context is in the prayer intoned every Sunday in all Christian churches. It begins with some pretty serious sweet talk, an appeal, a beseeching: "Our Father who art in heaven, hallowed be thy name . . ." (You are in heaven, and your name should be hallowed, sanctified, blessed, consecrated, revered, respected.) It continues, "Your kingdom come, your will be done, on Earth as it is in Heaven." (Not only are you hallowed but your kingdom, your will, your wishes, your goals are going to be carried out on Earth among us mortals as you carry them out in Heaven.)

At this point, if you are a parent ("Our Father") and your teenage son or daughter was intoning these phrases, you could be forgiven for thinking that they were leading up to "the big ask," a desire, a wish that only *you* could grant. But when the "you" is well, *God*, that might be a pretty big wish list. Everything from smiting your enemies, to riches, to personal beauty, to domination over the earth.

So it is jarring to hear the first request: "Give us this day, our daily bread."

They are asking—no, demanding ("give," not "please")—a plant. A specific one, wheat, that can be harvested, ground, and made into sustenance. And they are asking not for themselves but for all people: *our* daily bread.

In doing so, they are expressing an understanding of a fundamental relationship between body and spirit. Between physical needs and piety. You cannot experience God if you are hungry. It is difficult to practice compassion or charity or to ask forgiveness if you are without food. So in this prayer, we are asking God to help us show empathy and to deliver us from evil and temptation, and we are asking for—insisting on—food (bread) in return.

Gandhi understood this: "There are people in the world so hungry, that God cannot appear to them except in the form of bread."[13] Physical needs must be met before spiritual desires can be granted. And the best way to do this, the ultimate means, is through plants. Bread.

Nature is defined by plants, by luxuriant blossoms, arched boughs, colors that pulse in sunlight. The tallest living things, the biggest living things, the oldest living things are all plants (a coastal redwood more than 380 feet tall, a quaking aspen named Pando that weighs more than six thousand tons, and an unnamed bristlecone pine in California that may be more than five thousand years old, if you're interested).

All religions acknowledge and beg us to see God's presence in nature. "Nature always wears the colors of the spirit"; "I walk the path of Spirit[;] Nature is my sanctuary"; "In every walk with nature, one receives far more than he seeks"; and "Every flower is a soul blossoming in Nature" are all phrases that translate into heartfelt emotion when we walk through the woods, see a field of wildflowers, or walk barefoot on the grass on a summer's day.[14]

There is a spiritual commonality among plants, from flowers to trees, from spices to bread that will heal us, unite us, transcend religions, and provide a source of wonder for as long as human life exists.

Chapter Four

PLANTS ARE IMPORTANT

THE PART ABOUT WEEDS

PLANTS ARE so endemic to our surroundings that we can easily do a 180 and recognize another truism: not every plant is desirable. If we manage plants, as we must do in agriculture, then instituting our wishes on nature will, always and forever, go amiss.

Plants are in no way, shape, or form, static. Understandably, it is through shape and form that they compete with each other for limited resources. And if humans attempt to choose only a small handful of desirable plants to spread as a single uniform, monolithic monoculture, the soil will break open, cracks will ensue, and a new gang of tough plants will make an appearance (the poison ivies, the puncture vines). It remains a perpetual battle that has been waged for millennia, an ongoing struggle as to who's in charge: plants or humans?

For anyone who has attempted to manage plants, from the simplest of gardeners to the most sophisticated of giant, conglomerate landowners, the bulk of time spent doing plant management can be summed up in a single word: weeding.

And there is a powerful reason why: Weeds are and have always been the greatest limitation on the performance of desirable plants, those that are necessary for food, clothing, medicine, lumber, pasture, or sacrament. Not fungi, not bacteria, not viruses, not insects, but other plants.[1] Since the onset of human civilization following plant management, humans have been trying, working ceaselessly to manage weedy species as if their lives depended on it—because they do.

When we were a primarily agrarian society, with only a small set of urbanites, children were given summers off to help on the farm. One might ask why a first- or second-grader should have summers off. The simple answer? They had to help on the farm.

But doing what exactly? Six-year-olds can't drive tractors.

But they can use a hoe, a weeding implement that has changed little since first depicted in Egyptian hieroglyphics four thousand years ago. Indeed, prior to mechanization, before the plough, the size of a farmer's holding and their subsequent plant yields were determined by how well—and how fast—a family or group could hoe their land. The work involved was, and still is, such that some have suggested that weeding may symbolize the greatest labor achievement by human beings since civilization began.[2] And students still get summers off.

One would think, given the importance of weeds, that humankind must be pretty clear on which plants constitute "weeds." Well, yes and no. Determining what constitutes a weed is somewhat along the lines of how the Supreme Court defines pornography: You know it when you see it. A weed is considered an undesirable plant, sure. But the meaning of "undesirable" fluctuates in space and time. For example, a corn plant may be desirable if you want to grow corn, but the following year, if you are growing soybeans and a "volunteer" corn plant emerges, it will be considered a weed.

Such an approach, as you can imagine, is full of arbitrary definitions and delineations. "Weed" will vary in time and space, by culture, by environment, by need. Human attempts to scientifically characterize weeds from a taxonomic standpoint have failed. A weed in one culture may be an unloved flower in another.

Yet nature distinguishes weeds in ecological terms. They are renowned for their ability to exploit disturbance, their ability to colonize disorder. Such disorder can be raw and crude: floods and fires, earthquakes and hurricanes, events that wreck the existing plant order and reduce the mighty forests of oak, elm, pine, or *Symphonia globulifera* (chewstick, a forest species) to shards and pieces. But events that also offer opportunities for new colonization and domination—by weeds.

Weeds are designed for such circumstances. They can produce thousands, even hundreds of thousands, of seeds, seeds that can float, fly, and flutter to the four corners of the globe. Seeds that find a home in fresh soil or the aftermath of disaster, the perfect spot to germinate, to grow quickly, to impressively reproduce without any insect advisers, to dominate. They are, in ecological terms, pioneer species.

And they are tough. Weeds can take over spaces that you thought were unlivable and turn them into ecosystems. If you doubt me, go to an abandoned parking lot paved with asphalt, run over thousands of times by cars, sprayed with noxious chemicals. Over time, cracks will form, weeds will develop, from simple grasses to morning glory vines to weedy trees like Ailanthus, their roots splitting open the underlying human materials, depositing new seeds, new growth, new green. And after ten years, you might recognize that this river of tar has become a new plant home.

And then the pioneering weeds retire. Because the larger seeds from the nearby plant community aren't going anywhere. They may have been pushed aside following the cataclysm, but they will slowly, inevitably, start to return. They will once again tower over their pioneer cousins.

But humans have short-circuited the game. In managing nature for certain plants—ones that benefit only them—humans have become a much more rapid force for disturbance, removing native trees, ploughing the soil, and creating new environmental debacles. And pioneer species, weeds, preadapted to such upheavals are selected. Chosen.

Not only for agriculture: Weeds flourish wherever and whatever humans disturb. And disturbance in turn formulates its own unique set of weeds that directly threaten human use of that land, creating a perpetual land war. Rangelands, used as grasslands and scrubland for grazing livestock, are home to weeds that are poisonous or threaten grazing use. These include leafy spurge and yellow starthistle, cheatgrass and smutgrass, goatgrass and bull thistle, which are just a few of the more than three hundred weed species threatening rangelands.[3]

Forests have their own set of weedy warriors. James Miller of the U.S. Forest Service has identified more than thirty-three plants or groups of plants that are spreading rapidly through forests in the southern United States.[4] These include vines, which twirl and entangle, capture, and shade their freestanding brethren, including kudzu, English ivy, Japanese honeysuckle, and Oriental bittersweet, as well as other perennial weeds such as privet, wild rose, and bamboo and unwanted tree species such as the princess tree and tree of heaven. Then there are the riparian areas where water comes and goes, there when it rains, then siphoned away for irrigation or homeowner needs. Here the weeds, the pioneer species, have landed with an almost audible thud: purple loosestrife, curly dock, giant reed (Arundo), and salt cedar. They replace the native species, disrupt and destroy threatened and endangered native species, negatively impact recreational opportunities, and reduce water quality.[5] Not to be forgotten are the aquatic weeds, those adapted to living by floating directly on the water's surface. Ones whose growth is stimulated by the runoff of fertilizers used to encourage the growth of wanted agricultural plants: algae, hydrilla, water hyacinth. These weeds add a foul taste and odor to drinking water, stunt and kill off fish populations, restrict or prevent boating and recreation, prevent irrigation system flows, prevent commercial navigation in waterways, and provide habitats for disease vectors such as mosquitoes.[6] And it goes on . . .

We try to detect, to dictate, to determine the spread of these unwanted plants. To control them. But we have moved beyond the plough, the hoe, and the rake. To provide food for billions, we must kill unwanted plants by the billions.

As with any war, new strategies and new innovations are key. Yet for one brief, brilliant moment, it seemed our problem might have been solved. That moment relates to chemistry and the widespread use of herbicides: chemicals designed to kill unwanted plants.

If you have ever spent the day pulling weeds, standing in the hot sun, sweat pouring, rashes rising, you might be sympathetic to a quick and easy means of ridding yourself of unwanted plants, pioneer or not, environmental sensitivity be damned.

And chemical control as a quick, easy means to do so is not a new idea. For more than a hundred years, various chemical tactics have been attempted, including sea salt, sulfates, and nitrates of copper (sulfuric acid). Sodium arsenite for a brief time reigned as the most popular, being used on wide stretches of land from railroad rights-of-way to rubber plantations.[7]

There was just one slight, teeny-tiny issue. The side effects of sodium arsenite include skin irritation, burns, and loss of pigment; loss of appetite, stomach pain, nausea, and vomiting; headaches; and convulsions. (Oh, and it's also carcinogenic.)

But this was a prelude. Arising from chemical research during World War II, the new and improved age of weed killers began. From 1945 to 1965, more than one hundred new substances were developed, tested, and approved, including two key chemicals: 2,4-D (2,4-dichlorophenoxyacetic acid) and 2,4,5-T (2,4,5-trichlorophenoxyacetic acid). Their toxicity was such that even small amounts (one to two kilos per hectare, or two to five pounds per acre) were sufficient to control weeds.[8]

There can be side effects to the use of such chemicals, of course, serious ones, especially when you treat an entire ecosystem as weeds. During the Vietnam War, as part of Operation

Ranch Hand, millions of gallons of Agent Orange (a herbicide that killed any plant it touched) were sprayed over huge swaths of countryside. Why? Well, the U.S. military considered it a "two-fer": defoliating the rainforests meant less cover for the Viet Cong and destroyed their food supplies. More than twenty thousand square kilometers of forests and mangroves were destroyed over a ten-year period. And as it turned out, most of the crops killed were meant for the local population. Not an effective way to win "hearts and minds."

One other teensy issue. Agent Orange contained dioxin, a generic class of environmentally persistent compounds linked to everything from abnormal child development to reproductive damage to immune system suppression to hormone interference. And cancer.

Oops.

But such is their ease of use, relative to the grueling, sweat-inducing physical labor of weed removal that chemicals will continue to be used in agriculture. As farm size increased during the twentieth century, as mechanization and monoculture came to dominate, chemical pest control was a temptation too seductive to resist.

And perhaps the pinnacle, the zenith, of achievement was the discovery and implementation of one particular chemical, glyphosate. Glyphosate acts as a dummy molecule in a biochemical pathway unique to plants and in doing so disrupts protein production. It is highly effective: a couple of quarts per acre will kill even the toughest plants.

But it had one drawback: It didn't discriminate. It would kill any plant, which is good for a field of weeds but not so good for a field of crops and weeds. Solution? Well, you could genetically engineer crops to be resistant to this new chemical. Hmmm . . . if that were to happen, you could apply glyphosate to everything: millions of acres of soy and cotton and corn and . . .

So the company that owned the rights to glyphosate, Monsanto, did just that. In 1996, the first genetically modified soybeans, sold under the moniker "Roundup Ready," came on the market in the United States. ("Roundup" is the brand name of glyphosate.) By 1997, 17 percent of all domestic soybean acres transitioned to Roundup Ready; by 2001, it was 68 percent; by 2010, 90 percent. Such unequivocal adoption and use spurred additional innovation, and Roundup Ready corn and cotton swiftly followed.[9]

And here at last was the shining light, the last word: After the use of this chemical, weeds were done, finished, kaput. One and done. So successful was Roundup that after these genetically modified crops were introduced, a number of research articles were published reckoning that other management methods such as crop rotation and the use of other herbicides were not as effective.[10] Farmers did not need much persuading, and other forms of management, often much more inconvenient, were abandoned. Monsanto pressed its advantage, deriving "terminator seeds," a means whereby farmers couldn't use existing genetically modified seeds for replanting but had to buy new seeds each year.[11] Very soon, glyphosate dominated the weed killer market. As for evolution, some mutation that could render the resistance futile? Monsanto assured their users that two mutations were necessary for that to happen and that it was very unlikely, implausible, improbable.

They might have been right. But as the adage goes, "They were victims of their own success." Normally mutations, especially double mutations, take time. But glyphosate wasn't just dribbled onto the land, it was poured on. Ten years after glyphosate was introduced, *more than 150 million pounds* were being used in the United States alone, and this just in agricultural (i.e., not homeowner use). As of 2019, that figure is between 250 to 300 million pounds.[12]

Use rates in the hundreds of millions of pounds is resistance selection on steroids. By 1998, resistance to glyphosate was observed in Italian ryegrass. This was followed by resistant horse weed in 2001. Today, twenty-four weed species are confirmed resistant, fourteen in North America.[13] Glyphosate resistance is prevalent in America's heartland, with some resistant weeds such as Palmer amaranth growing large enough to stop combines in their tracks.

This is not the whole story of course; glyphosate may have ridden the chemical wave, but it was not alone. The Weed Science Society of America reports that at present, herbicide-resistant weeds have been reported in ninety-two crops in seventy countries and that weeds have evolved resistance to twenty-three of the known herbicide sites of action. Multiple resistance to different herbicides has now been confirmed in more than one hundred weed species.[14]

If the war against weeds seemed to have been won with glyphosate, it was an illusion, but one with consequences. Companies are back to the drawing board, or wet lab, hoping to find new solutions. In the meantime, old and new combinations of chemicals are being thrown together, and new genetically modified crops resistant to these combined mixes are being generated. Oh, and Roundup is still being sprayed. But because of the increased resistance of weeds, greater and greater concentrations are being used.[15]

However, the lure of achieving chemical control over weeds still remains. If glyphosate was once a cure-all, it is now a crutch. Where once a single spraying per year was sufficient, now three to four are necessary. It is ironic that at this point in the struggle, hundreds of millions of pounds of herbicides are being used that wouldn't have been if glyphosate had not been invented.

And so it goes. The war continues. Humans, certain that they can conquer plants, choose and abuse what they need when they need it. And to some extent, they can. It is no coincidence that the world manages enough crop plants to support almost eight billion people.

But the human species would be very stupid to think that plants are without the ability to respond, without resources. The war against weeds is in its own way illustrative of progress and regression, of endless toil, of technological innovations, of chemical triumphs, and of the mutations that negate them. We would do well to remember that plants are by far the dominant kingdom.

Chapter Five

PLANTS ARE IMPORTANT

THE PART ABOUT ART—AND ALLERGIES

PLANTS CONTRIBUTE to human activities in ways that cannot be completely compartmentalized or instantly comprehended.

Art and music are exemplary of the human condition, and plants are, and will remain, a source of inspiration. Any art museum tour will include paintings of trees, grass, and flowers, and plants have served as muses for artists since trees were etched onto ancient cave walls. Just think of Van Gogh's mastery of *Irises*, *Bouquet of Flowers* by Manet, or *Jimson Weed/White Flower No. 1* by Georgia O'Keeffe. Music, too, is inspired by plants; for example, Tchaikovsky's *Waltz of the Flowers*, Puccini's string quartet *Crisantemi* (inspired by chrysanthemums), *Hyacinth House* by The Doors, and *Wildflowers* by Tom Petty. As for the written word? Here the context is replete with examples, from Robert Frost's *Stopping by Woods on a Snowy Evening* to Joyce Kilmer's *Trees* to Michael Pollan's *The Botany of Desire*.

But plants are not only inspirational. They provide support, functionality. Sure, stopping by woods one snowing evening can precipitate questions about God, but woods also provide paper: paper to write on, to compose poems, to hold the colors and images drawn from your imagination. Trees also provide the material needed to generate music, from guitars to violins to pipe organs. Wood supplies basic architectural materials for new designs, new shapes, new spires, new highs.

That functionality, that support, is neither limited to the arts nor constrained to humans; it carries through to all living things. Plants provide the most basic of necessities. Plants provide oxygen, an essential gas for all movement. Plants lift water from the soil to the atmosphere via transpiration, water moving from soil to roots, from stems to leaves. Entire forests lifting water, creating their own clouds, their own rain. Roots sifting, filtering out contaminants, purifying the water before it enters streams and lakes—one reason why planting trees around waterways can restore lakes and streams. Plants can, in an almost magical fashion, convert the basic essentials of air and water, soil and sunlight into sustenance for all living creatures. They are chemists of the highest sophistication, creators of the new and unimaginable, suppliers of animals' wants and needs. They provide homes, shelter, shade, and structure. They make life possible for all creatures, big and small, shrimp and whales, butterflies and elephants. Plants are the dominant life form on Earth, and it is only through their munificence that we exist.

But we humans suffer from a peculiar form of blindness, of arrogance. We do not see plants as special, as the bedrock of life, as the basis of civilization. They are, for almost all of us, a green curtain, a background, a canvas that is constant. Yet if we move the curtain aside, we see the wizardry: nectar for bees, sugar for hummingbirds, grass for bison. Krill, which feed on phytoplankton, a primitive plant form, in turn feeds the largest of mammals, the blue whale.

Even behind the curtain, some events are not immediately visible. Belowground, root growth is promoting and sustaining soil-borne diversity, with a teaspoon of topsoil containing six billion microorganisms, all dependent on plant roots for nourishment.[1] Soil is a "microverse," containing everything from fungi to earthworms, a microverse essential for the recycling of all earthly things. Treetop canopies, invisible to the human eye, act as animal high-rises and are home to 90 percent of the animals in the rainforest. They are filled with insects, birds, monkeys, frogs, sloths, and even other plants that manage to find a niche among the lofty limbs.[2]

But before I climb too high into the laurel tree of plant praise, let me point to the other side of the coin. Musicians, especially those who play a woodwind instrument, likely know about arundo, a large bushy plant found in riparian regions around the world. They are familiar with this plant because the hollow stems of arundo, called culms, are used to make reeds, a necessity for many woodwind instruments. And so far, no satisfactory substitute has been found. Arundo has played a significant part in Western culture, influencing the development and spread of music. But if you study weeds, chances are you've also heard about *Arundo donax*, or giant reed. This plant is considered an outside threat to native ecology, as it can invade and dominate wetlands, including stream banks and lakeshores. It degrades wildlife habitats, interferes with flood control, and increases the risk of fire.[3]

Or take another lovely plant, one that floats placidly on the waters of the world, native to the Amazon basin, its flowers delicate shades of pink or lavender. It is exquisitely beautiful, to the extent that it was widely collected and sown all over the world. That plant is the water hyacinth, and its spread—helped by human appreciation of its aesthetic beauty—has turned a lovely flower into a bellicose monster, a plague that threatens the biodiversity of lakes and rivers, impedes the flow of water to homes, and chokes canals meant to deliver water to crops. Its fearsome ability to resist human attempts to control its spread is legendary and has elevated the water hyacinth to many lists of the world's most damaging invasive species. How invasive is it? It was first sighted on Lake Victoria, Africa's largest lake, in 1989, and by 1995, 90 percent of the Ugandan side was covered with it.[4]

Plants are integral, essential to life. No ecosystem in the universe can exist without organisms (plants) that can convert solar energy into chemical being. Yes, plants are beautiful, even inspirational, but they are also basic and pragmatic essentials that reflect life's functionality. We can gaze at a flower, moved beyond words by its aesthetic beauty, but then twitch our nose and sneeze because of the pollen emitted from that flower.

We have grown complacent, so conceited in our beliefs that we are superior and that we can control where and when plants grow. We are blind to the consequences. This is an endemic green shade of disorientation.

In Part I, I have tried to communicate why plant biology is so critical, so essential to human existence. To convey not only their beauty, but their consequence in life as we know it. I did this in the hope that it would foster appreciation for their function.

And to ask the next obvious question: If plants are important in the ways described here, what would happen if humans altered plant biology? All of it, all over the globe. Human tinkering that would impact everything from plant chemistry, species diversity, and agricultural production to the cultural, social, and medical aspects of our relationships with plants. All the important stuff.

PART II

PLANTS AND MAGIC

MY GRANDFATHER and uncle were both magicians, at least part time. As a child, I was mesmerized by their antics to produce something from nothing with an elaborate twisting of the hands, a finger-flick, followed by a smile and a nod of the head. I would watch them, eyes wide, mouth open. I wanted to know more.

Even when they showed me how the trick was done, I was mesmerized. As I tried to copy them, I learned to appreciate the skill, practice, and patience involved in producing that something-from-nothing moment.

Gardening may seem like a subject as far removed from magic as fish from bicycles, but for me at least, it produced the same effect: something from nothing. Toss a few corn or tomato seeds into a pot, and something remarkable and joyous happened. Plants materialized, growing stately and tall, green and natural. It felt as satisfying as any complicated magic, and much, much simpler.

And something of use! Something that was appreciated in a working-class home: food. Unlike any product from magical conjuring, it was food that could be eaten. And was. As I grew older, I grew bolder. I added flowers and trees to my garden. And something else grew inside me, a sense of wonder, an appreciation for living things. I was part of something, part of nature. The world flowed through me—just a little.

This is not to say that nature is, under all circumstances, wonderful. I've stepped barefoot on puncture vine and scratched my forearms on rosebushes to the point where I resembled a human pincushion.

But the magic, the childlike wonder of plants, whether they be flowers or food, vines or velvetleaf, has never left me. In part I suspect, because plants were not an illusion; they were powerful, real, consequential.

And I wanted to know more.

Chapter Six

SCIENCE IS FUNDAMENTAL

IN PART I, I tried to translate some of the magic of plants into words. To provide a glimpse into their green and fascinating cosmology. But part of magic is seeing how a trick works. Appreciating the fundamentals and the practice, knowing the angle, the gaff, the effect of patter to produce the trick. I know for some, finding out how a trick works destroys the illusion. But to my mind, it strengthens it. It provides value, insight, knowledge. It satisfies.

And for me, plants, the green "magic" arising from the earth, were reason enough to study the mechanics behind the visualization. So, on to science. Plant science. And soil science. And environmental science. Seemingly simple; after all, how complicated can plants be? Well, agronomy, aquaculture, botany, ecology, entomology, forestry, genetics, horticulture, pathology, physiology, pomology, oenology, toxicology, and viticulture are just some of the subdisciplines of plant science.

It was pretty damn fascinating. As often happens, the more you know, the more questions you have, and the motivation to dive deeper, to find connections, to seek understanding and insight. I was fortunate to have mentors along the way who encouraged me and allowed me to explore my questions.

When doing science, scientists come up with a question that we think is interesting, one that we want to answer. We attach a hypothesis to that question, we test the hypothesis, we describe how we tested it (in enough detail that others can repeat what we did), we present the results, and we discuss those results in the context of earlier findings. We then send this document to a journal where other experts (our peers) can review what we've done for the soundness of our hypothesis, the veracity of our results, and the possible impact of our findings on the "bigger picture."

I cannot speak for all scientists, but I have *never* gotten a "rubber stamp" on any study I have written up and submitted to a journal for publication. Typically, I get a laundry list of questions of the "Have you thought about X or considered Y?" variety. This peer-review process, which provides the opportunity to answer important questions, is the core of the scientific method. Trust me, if there is one human facet endemic to all scientists it is this: We are not shy about pointing out other scientists' mistakes.

If you think that the reason for the increase in atmospheric CO_2 has nothing to do with human activities and won't affect plant biology, please challenge the data presented. Not in a hostile way but using the scientific method. Formulate a hypothesis for why atmospheric CO_2 is increasing *without* invoking human activities such as the burning of fossil fuels. Show that plants do not respond physiologically to more CO_2, and prove your hypothesis through experimentation. Tell us how you did it, so we can replicate the results. Let other experts evaluate what you did for possible errors or bias.

The data I will present have been through that process—peer review—but that doesn't mean they are the last word—far from it. Unlike politics or religion, science encourages questions and doubts. To the point that scientists want you to test your questions and publish the results. So if you spot a consistent error, something we missed, please tell us.

When it comes to plant science, there are a couple of basics that you probably remember from your elementary school days. Did you ever plant a bean inside a wet paper towel with its ends dipped into water at the bottom of a glass jar, and then place that jar on the windowsill? And when the seed broke open, the green shoot emerged, and leaves grew above the glass rim, did your teacher ask, "What do plants need to grow?" If so, you probably remember the answer: light, water, and nutrients like nitrogen and phosphorous.

Now let's engage in a little experiment. Let's suppose (a wonderful word, "suppose"). Suppose you had a superpower: the power to increase a resource plants need. Not just in your backyard, your neighborhood, your city, your town, your county, or even your state. But in the entire *world*—every plant in every ecosystem. And in increasing this resource, you would of course alter all animal life—and all bacterial life—from aardvarks to zebras, from ants to zorapterans, as they all depend on plants for food.

Let's be more specific. Let's say you've already increased this global resource by 25 percent and that you're planning on increasing it by another 50 percent—in your children's lifetime. That's a pretty awesome power. Is that even possible?

OK, we know plants need light, and yeah, humans can scatter a few grow lights hither and yon, but no way are we controlling some colossal dimmer switch on the sun. So that's out.

And they need water. And while we do some irrigation and dampen the occasional houseplant, no way we have already increased rainfall—globally—by 25 percent.

How about nutrients? Well, this might be a bit more possible to achieve. After all, about 40 percent of all land is used for agriculture, and we add fertilizer, nitrogen, phosphorous, potassium, etc. But still, our efforts fall short of a global goal of a 25 percent increase.

How about carbon dioxide?

A little background. You and I, your mom and dad, your cats and dogs, your goldfish and hamsters—all living things—consist mostly of carbon. Carbon is a primary component of our bodies: our protein, our lipids, our nucleic acids (DNA and RNA), and so on. Plants use light to split water molecules and use that energy (hydrogen power!) to capture carbon—carbon necessary for all life's chemistry—while adding the oxygen from the split molecule to the air. Hence, photosynthesis: taking light (*photo-*) and turning it into carbon (*synthesis*). Of course, you have to add some other nutrients, such as nitrogen to make protein or phosphorus to make DNA, but carbon is literally the chain that holds all life together. But where does that carbon come from?

From carbon dioxide in the air. And therein lies a problem. Once upon a time, one to two and a half billion years ago (even before the dinosaurs), there was quite a bit of CO_2 in the air, about ten to two hundred times more than today. (If you wonder why the earth didn't melt with all that CO_2 in the air, the answer is that the solar output at that time was considerably less than it is today.)[1]

But for the last few million years, owing to various weather events, CO_2 in the air hasn't been that high, shifting between 200 and 300 ppm (parts per million).[2] At that level, plants are literally starving for carbon—photosynthesis is not operating at full capacity. And since photosynthesis provides the carbon necessary for plant growth, plants aren't responding as much as they could.

So back to basics. CO_2 is a resource. If I increase it, plants will respond more (this fact is based on the findings of hundreds of studies).[3] More CO_2 equals more plant growth.

But wait, how do humans increase CO_2? Well, we burn fossil fuels (coal, gas, oil, etc.). All of which are—wait for it—dead plants (and a few animals). Plants that died over millions of years, converting CO_2 into carbon-rich biological material. So when I burn fossil (carbon) fuels, I oxidize them; simply put, oxygen (O_2) and carbon (C) make CO_2. It may have taken millions of years to store all that carbon, but we are using it up in the blink of an eye, geologically speaking.

According to the CO_2 monitor at the Mauna Loa Observatory in Hawaii, ambient background CO_2 (from car tailpipes, people breathing heavily, etc.) has increased by about 25 percent since 1975,[4] as figure 6.1 illustrates. And it is estimated

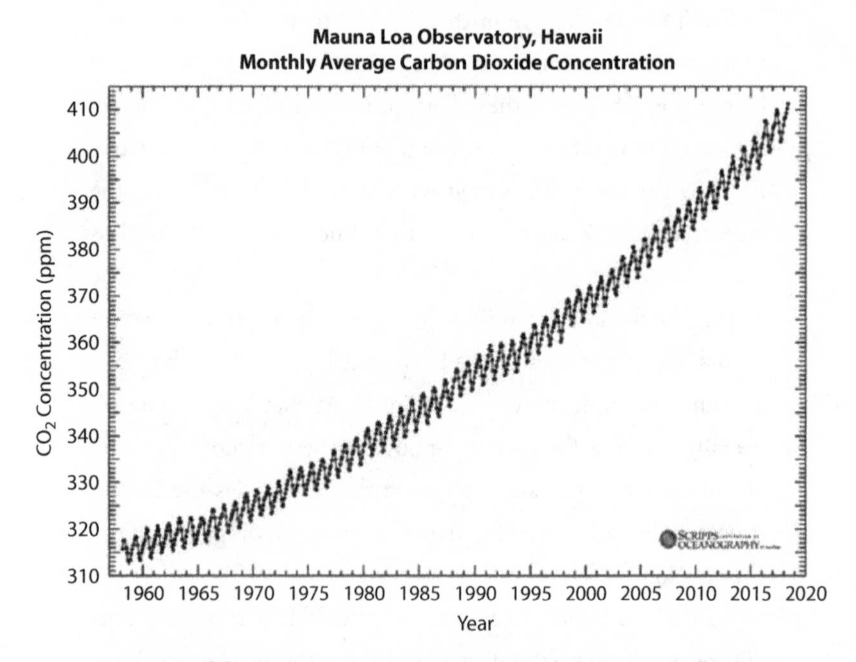

FIGURE 6.1 The Keeling curve, which has been keeping track of background CO_2 since 1957. Data from Scripps CO_2 Program, last updated June 2018.

National Oceanic and Atmospheric Administration, public domain.

to increase by another 50 percent to between 600 and 800 ppm by the end of the current century. (If you're curious, those little wiggles reflect the summer and winter uptake of CO_2 by vegetation—trees in the northern hemisphere are bare in the winter and thus take up more CO_2 in the spring and summer).

So we have added, and will continue to add, CO_2 to the atmosphere—a resource that plants need to grow. Which brings us to an obvious question: Why is this a bad thing? "CO_2 is plant food," remember? I suppose to some extent, it depends on your perspective and your previous plant experience.

From a nonscience point of view: Meh.

From a botanical nerd point of view: How will this impact the growth of my tomatoes? The availability of my tea? Or my coffee? What about my Wheaties? What about my marijuana plant? (Cough, cough.) What about my cotton dungarees? What about my orchids? What about the hypericin from the Saint-John's-wort I take for my depression? What about the reeds for my clarinet? What about the weeds in my garden? Oh, man, what about my kudzu problem?

From a plant scientist point of view: Will all plants respond in the same way to this increase? To future increases? If not, how will it affect species diversity? Will some plants be "winners" and others "losers"? How will this affect the global food chain? How will it affect insect fecundity? Weeds in agriculture? Will food sources for disease-carriers like mites and mosquitoes increase or decrease? How will Indigenous peoples who rely on plant-based medicines be affected? What about the global narcotics problem? Will *Erythroxylum coca* (cocaine) production be affected? What about herbicides like Roundup? Will it affect chemical weed control? Will CO_2-induced growth stimulation affect the nutritional quality of the food I eat? What about food for other animals? What about invasive plant species like puncture vine? Or parthenium?

From a conservative politician's point of view: CO_2 is plant food. Plants are good. Damn liberals, don't they recognize that "we are living in a lush environment of plants and animals as a result of the carbon dioxide increase? This is a wonderful and unexpected gift from the industrial revolution."[5]

From a liberal politician's point of view: No. Any positive effect of increasing CO_2 on plants will be overridden by an increase in fires and storms and rising sea levels and temperatures. And even if there were a positive CO_2 effect, it would be, uh, minimal. Or dependent on something like nutrients or pests. So, no beneficial CO_2 effect. Ignore. Focus on climate change. Starving polar bears, sea levels rising.

■ ■ ■

So it goes. Most people at least recognize that CO_2 makes plants grow. And that atmospheric CO_2 is increasing. But what this ultimately means is unfortunately dependent on what you *want* to see. If we want to know the reality, we need to know more.

In the pages that follow, let's take a closer look and try as best we can to take a scientific lens to what the recent and projected increases in CO_2 will mean for plant biology and for all of the living things that rely on plants, including us.

There are few disciplines, few career choices where confrontation and questioning are encouraged. Science is the exception. So using that science lens, I will dive deeper into the "CO_2 is plant food" issue. The issue is multilayered, complex, and full of twists and turns. But along the way, I will do my best to wave my hands and apply science to explain the magic behind plant biology. First to illustrate the good—to see what more CO_2 could mean for plants that benefit humankind. Then to illustrate the bad—to examine the negative impact of more CO_2 on plant biology and ecology. And finally to illustrate the "OMG"—to look at what rising CO_2 means for plant biology—and for life as we know it.

Chapter Seven

CO_2 IS PLANT FOOD

THE GOOD

PEOPLE FOOD

Right. CO_2 is plant food. And we eat plants, mostly rice, wheat, and corn (which make up about half the global consumption of calories; with the addition of a few more cereals like sorghum and a few more root crops like potato, that's close to 80 percent of the global food supply). So more CO_2, more food. Yes?

Maybe. The numbers vary, reflecting the methodology by which additional CO_2 is calculated (e.g., indoor vs. outdoor experiments), but for now let's look at outdoor field experiments for three major crops: wheat, rice, and soybeans.[1]

Wheat: A future CO_2 concentration of 700 ppm (expected by the end of the century) will stimulate wheat yields by 19 to 26 percent.

Rice: At the same future CO_2 concentration, rice yields will increase by 13 to 26 percent.

Soybeans: A 700 ppm CO_2 concentration will increase the yield of soybeans by about 28 to 40 percent.

Well, that's the future; what about recent CO_2 increases? (The concentration of atmospheric CO_2 has increased by 25 percent since 1975.) It's a bit spottier here, in part because it is hard as hell to remove CO_2 from the air—which you would need to do if you wanted to simulate prior CO_2 levels. To make matters even more complicated, how do you account for new and improved crop varieties, or the large increase in fertilizer application, or the new equipment that has come into play in recent decades? Although difficult to work out, the following estimate has been established for wheat: an increase in yields of 19 to 48 percent with a CO_2 concentration of roughly 275 to 360 ppm.[2] Or, according to my estimate, which is a bit higher: an increase in yields of about 50 percent with a CO_2 concentration of 293 to 385 ppm.[3]

As for outdoor experiments on rice or soybeans, there is nothing yet. However, scientists are looking at comparing yield trials for wheat, rice, and soybeans from the 1960s (i.e., when atmospheric CO_2 was about 320 ppm) with yield trials from 2014 to 2019 (when atmospheric CO_2 was about 410 ppm) using the same cultivars, location (i.e., soil), and management techniques, among other factors. We expect to find a stimulation in growth and yield. This is pretty cool, right? Recent and projected CO_2 increases can stimulate the growth and yield of some basic cereals.

Wait, there's more. Note the variation in the response of yield to recent (wheat) or future (wheat, soybean, and rice) CO_2 increases (19 to 40 percent among all wheat varieties). Could we exploit that variation? Could we select for varieties that can convert more CO_2 into more seed?

To put it mildly, there is good reason to do so. At the moment, it is clear that the number of individuals who suffer from food insecurity (not knowing where your next meal is coming from) is rising.[4] The number of undernourished people in the world has been increasing since 2014, reaching an estimated 821 million in 2017. The COVID-19 pandemic is expected to add to that number, with an estimated 845 million people being food insecure in 2020.[5]

So how much variation in response to rising CO_2 is there among cereal varieties? The largest comparison to date has been with seventeen cultivated rice varieties conducted at the International Rice Research Institute in the Philippines.[6] That study, done indoors, indicated that there was significant genetic variation in yield at future CO_2 concentrations of about 650 ppm. Cool.

And there is evidence of significant yield variation within a number of important crop species, including common bean, cowpea (black-eyed pea), peanut, rice, soybean, and wheat.[7] Enough variation, in fact, for breeders to start selecting for CO_2 responsiveness to future CO_2 levels. Indeed, although there are fewer data, breeders may even be able to ascertain variation among crop lines to the change in CO_2 that has already occurred.[8]

Sounds easy, but it's actually a little more complicated. Increasing the CO_2 concentration by doubling it all at once can provide interesting but unclear results. A better outcome would occur if you evaluated crop varieties or populations by gradually increasing the CO_2 concentration over many seasons to simulate real-life conditions. Unfortunately, there are almost no long-term studies along these lines. So at the moment, we don't really know for certain which crop varieties can convert more CO_2 into more seed yield. Sad, right?

As always, along the fringes, there are a few interesting exceptions. Not for crops but for a grass—*Dactylis glomerata*, or orchard grass—which is grown around the world for hay and forage. A grass that is fortunately found along natural CO_2 springs.[9]

Scientists have examined populations of this grass from both ambient- and high-CO_2 locations (from a CO_2 spring) and found that one of the main selection factors was more CO_2.[10] In fact, when populations from each CO_2 environment were tested, it turned out that the population adapted to high CO_2 produced 30 percent more biomass relative to the population adapted to ambient CO_2. This outcome suggests that over time, plants can respond to and exploit additional CO_2 for growth.

To sum up: There is a ton of evidence indicating that different crop varieties will respond differently to future CO_2 increases (there are fewer data for recent CO_2 changes), and breeders could use knowledge of these differences to select for CO_2-responsive crops. Such efforts—if CO_2 springs are any example—could result in better yields for agriculture and enhanced food security.

So why isn't anybody doing this?

BECAUSE PLANT BREEDERS ALREADY ARE?

I've raised the breeding issue with agribusiness leaders and scientists on several occasions. And the response is always the same: a condescending smile, a pat on the head, and the assurance that "we are already doing this." (OK, they didn't really pat my head, but the condescending smile was real.)

So when I impertinently replied that to my knowledge, there are no published records of any methodical, long-term attempts on the part of breeders to select for crop varieties with greater yield responsiveness to rising levels of CO_2, I was told, "Well, of course, they are doing it automatically."

When I pressed for details, I heard the following, "Breeders always select for improved yields, right? So as CO_2 increases, those increases will be incorporated into trials for new varieties." Science is, at its core, a weird mixture of skepticism and wonder (not unlike watching a magic show). This time my skeptical side was stirred, and I decided to see if these claims were true.

Where to begin? My hypothesis ("Suppose!") was pretty straightforward. If breeders are already selecting for CO_2 sensitivity de facto (a fancy Latin phrase meaning "even if we don't know what we're doing, we're doing it"), then modern, adapted CO_2 lines should show a greater response to CO_2 relative to varieties that were developed earlier, say during the early twentieth century. That greater sensitivity would be a result of inadvertent selection over time.

With the help of a supportive and very bright colleague, Dana Blumenthal, a USDA scientist, we tested this hypothesis by looking at oat varieties from the 1920s with related oat varieties released during the 1990s. Or, in CO_2 terms, we compared oat lines released when atmospheric CO_2 was about 300 ppm (in the 1920s) with those released when atmospheric CO_2 was about 360 ppm (in the 1990s)—an overall increase in CO_2 of about 20 percent. We compared lines from the same location, from the same soil, and from the same physical environment. And because oat doesn't cross-pollinate (well, at very low levels), it was unlikely that the lines had not changed genetically since their initial release.[11]

If breeders are already choosing the most CO_2-sensitive lines, we didn't see it. In fact, it was the oat varieties released during the 1920s that demonstrated a greater relative and more diverse response to the twentieth-century increase in atmospheric CO_2. We repeated this study for wheat lines released early and late in the twentieth century and observed a greater relative yield increase in response to increasing CO_2 for lines that had been released earlier.[12] Similarly, scientists in Japan recently examined CO_2 responsiveness among Japanese rice lines released since 1882 and found that breeding had not, in fact, increased the CO_2 responsiveness of rice.[13]

Three studies do not a scientific "law" make. Maybe breeders are passively increasing CO_2 sensitivity among crop lines. But as I write this, I still can't find any evidence that breeders have maximized CO_2 responsiveness among newer cereal varieties, even though atmospheric CO_2 has increased significantly during the twentieth and early twenty-first centuries.

But if passive selection were occurring, it would have to overcome some biological obstacles. First up is the primary enzyme all plants possess—the world's most abundant protein—and one essential to photosynthesis, as it is able to "capture" CO_2 in the air. It is called rubisco (rhymes with "Nabisco") and is short for "ribulose bisphosphate carboxylase oxygenase" (sorry). Passive selection isn't enough to change the concentration of this, the world's most fundamental enzyme. Indeed, some scientists have suggested that rubisco concentration is stuck at CO_2 concentrations that have existed for the last million years (i.e., when the global concentration of atmospheric CO_2 was between 200 and 300 ppm) and that optimizing rubisco (by genetically modifying it) would increase photosynthesis by 10 percent.[14]

A second obstacle is recognizing that plant breeding often has numerous objectives, not just an increase in yield. For example, if your goal is to come up with a variety that gets eaten less by locusts, your focus isn't on yield per se but on discovering chemical changes that might prevent a locust lunch.

And once that variety had been discovered, you would want to maintain it exactly as is for perpetuity. From a breeding standpoint, you would want to have that line available for further breeding—it would do zero good if you were to lose the original trait you had worked so hard to find. If I have a variety resistant to rust (a common fungus), I want to maintain that line in its purity. If I released a crop variety in 1980 that is copyrighted, I want to release the same variety in 2020. I don't want it to change genetically, because if it does, I can't get any money for it.

Another factor is also at play. We grow food today more or less as monocultures. Let's say you like McDonald's fries (I do).

But McDonald's relies primarily on a small subset of potato genetics, primarily those of Russets (with Russet Burbank being the most popular).[15] So if I'm a potato farmer, I grow Russets. Indeed, almost every current business model for food production relies on a small subset of existing genetics. Farming is a hard business; do I want to grow some new variety if no major company will buy it? (Nope.)

For sound business reasons, we have assumed the faster, more uniform, and cheaper assembly-line model of food production. A model that doesn't reflect the inherent variation in nature of a given crop but a narrow subset that meets economic needs. Hence, Russets.

Yet in narrowing my gene pool to achieve uniformity, I am also reducing the chance that I will find that genetic combination that will optimally respond to rising CO_2 levels (or other rapid environmental shifts). For example, there are more than one hundred thousand rice lines,[16] but only about a dozen are widely grown in the United States. Without active testing, what are the chances that breeders just happened to choose the most CO_2-responsive rice varieties? But a hundred years ago, when food production wasn't done via assembly line, finding diversity was the norm, and, at least theoretically, you would be more likely to find a gene combination that could respond to an environmental change like more CO_2. So far, the data we do have support this scenario. But ultimately, if I have a handful of varieties that represent the bulk of my business (and are proprietary to boot), why would I *want* to change?

CHOICES AND CHALLENGES

Globally, there is a widely acknowledged need to increase agricultural productivity to address food insecurity. Yet as we have seen, systematic attempts to exploit rising CO_2 to try to increase crop yields have been limited. I confess to being surprised by this. If levels of soil nitrogen had increased by 25 percent globally since 1975, wouldn't there be interest in selecting crop varieties that could exploit that increase to boost seed yield?

In fairness, one can recognize the challenges in selecting for CO_2 responsiveness. The most obvious is methodology. How do you simulate future CO_2 concentrations at the field level for an entire season? Not everyone has a natural CO_2 spring in their backyard.

The U.S. Department of Energy (DOE)[17] took up this challenge, modifying methodology from the 1980s: specifically, standpipes arranged in a ring with each standpipe directing gas into the center of the circle. Such a design was originally conceived for determining the impact of sulfur dioxide (SO_2), a pollutant emitted from coal-fired plants, a gas associated with the negative impact of acid rain on forests.[18]

The DOE discovered that such a design could also be used for delivering CO_2, and they came up with the catchy acronym FACE (standing for "Free-Air CO_2 Enrichment") to describe it. FACE rings could be put into fields, and evaluations of the responses of crops such as wheat, rice, and soybean to increasing levels of CO_2 could be made. Cool.

But not cheap. Adding more CO_2 to plants growing in a field requires tons of CO_2. Equipment, monitors, and people are also needed to make sure that the cloud of CO_2 stays in the center of the ring. Still, the process allows us to simulate future CO_2 levels in a real-life setting. And provides the opportunity to begin selecting crops for CO_2 responsiveness. As such, FACE became the gold standard for doing CO_2 studies, and trying to get a non-FACE study on elevated CO_2 into a high-profile journal was nigh on impossible. Trust me.

Perceptions change, however. Trying to maintain a consistent level of higher-than-normal CO_2 in a field when the wind is blowing ain't easy. In addition, to save money, most researchers gave the extra CO_2 only during the day. But it turns out that rapidly fluctuating CO_2 levels or not exposing plants to CO_2 at night might, in fact, underestimate any fertilization effect of increased CO_2 on plant growth.[19] Conversely, while enclosures like greenhouses could allow for consistent CO_2 delivery, they don't represent real-life conditions. So from a methodological point of view, setting up a large study in which hundreds of crop lines could be evaluated for CO_2 responsiveness is not so easy.

Hmm. Now what? Well, rather than try to look at the future, what about the past? Remember, CO_2 has already risen by 25 percent. What if we look at how that increase has already affected yield responses in the field?

Wait, that would be hard. As we've seen, CO_2 responsiveness varies with variety, and we aren't growing the lines of wheat, rice, and soybean today that we were in the 1960s.

But what if we did? What if we repeated yield trials using the same varieties, at the same locations, using the same management technique (e.g., row spacing) and compared our results with those from the 1950s and '60s? Julie Wolf (another USDA brainiac) and I are leading a team of scientists doing exactly that—analyzing soybean trials that began in Maryland in the 1940s and continue today. A multiyear trial to account for year-to-year variation.

What have we found so far? The studies are still underway. But if we can find some optimal CO_2 responsive lines, the next step will be to identify why they are optimal. Is it related to flowering? Leaf shape? Photosynthetic stimulation? Tillering? Can we pass that information on to breeders? Could rising CO_2 potentially spark a second green revolution? Are modern lines as CO_2 responsive as older lines?

Stay tuned. To my knowledge, this will be the first evaluation of lines in response to recent CO_2 increases. Something that, to date, no other group—with one notable exception—has done.

UP, UP, AND AWAY

Given that CO_2 is a resource and more of it can make plants grow bigger, there are a few commercial uses for increased CO_2. For example, nursery and floral production rely on greenhouses being supplied with additional CO_2 to make more flowers. But one of the more curious uses was evident in a visit I made to southern California.

About ten years ago, my colleague Dr. James (Jim) Bunce (my science mentor) and I were invited to visit a new company: AG Gas. We were shown a field of strawberries in

Ventura County, one that resembled lined paper, as shown in figure 7.1.

The little patches of green are strawberry plants, and the black tubing under the plastic is delivering CO_2. The CO_2 is released under the plastic and vents through holes positioned where the strawberry plants are growing.

And where did the CO_2 come from, you ask? As oil companies drill for oil, they often encounter gas burps: large quantities of CO_2 at the top of the oil pocket. Rather than venting the CO_2, AG Gas decided to try to come up with a

FIGURE 7.1 CO_2 fumigation of strawberries in Ventura County, California.
Author's photo.

means of making bigger strawberries—this from an oil company no less.

Was it successful? Jim and I tried to answer the company's questions about CO_2 and plant biology and discovered that they had two experimental fields: the one shown in figure 7.1, which doubled strawberry yields when the plants were given extra CO_2, and another site nearer the ocean, which was much windier. There, the wind kept the CO_2 from concentrating.

This result was interesting and proposed a promising means of increasing crop yields in the field. But such a setup is expensive, and as times changed, so did AG Gas. To pay for the extra CO_2, you need a crop that commands a good price—like fresh strawberries.

Or cannabis. And that is what AG Gas is now working on: using CO_2 to enhance the growth of cannabis in hoop houses where strawberries once grew. If you check out their website, you will note the claims that increased CO_2 increases yields by 20 percent or more and that bud density increases, as does floral quality and water use efficiency.[20]

So how unusual is it that a business is using CO_2 to stimulate the growth of marijuana? Go to Google and search "CO_2" and "marijuana." I'll wait.

Was the first hit something like "How to use CO_2 to make marijuana grow more?" Ironic, isn't it? Of all the crop industries that could use CO_2 to increase yields, the pot growers are the global leaders.

Where can I find a source of CO_2 for my cannabis? There's no shortage of information online, from simple high school science fair tricks to sophisticated pressure regulators. When should I apply CO_2 to my cannabis? Again, you'll find lots of information, with an emphasis on appropriate temperature and light levels. How can an increase in CO_2 affect THC concentration? You'll find no scientific studies per se but apocryphal suggestions that additional CO_2 increases THC levels. What marijuana variety responds best to increased CO_2? Off-the-record conversations indicate that certain varieties respond better than others, but there has been no clear scientific evaluation of this. I have read that agriculture began in part because hunter-gatherers wanted to grow certain plants that made the world a little easier to live in. Wouldn't it be nice if agribusiness learned from the pot growers? History repeating itself.

WAIT, THERE'S MORE!

If adding CO_2 can help you get high, there are numerous folks who would lump that in the "more CO_2 is good" category—but not everyone. Are there other aspects that can be seen as "good"?

Let's consider malaria for a moment. Yes, malaria. An infectious disease caused by the introduction of a *Plasmodium* parasite to the human bloodstream by infected female *Anopheles* mosquitoes. It is recognized globally as a scourge—a disease that infects more than two hundred million people each year.

Within two weeks of being infected, people experience shakes, chills and fever, nausea and vomiting, and headaches. Malaria kills one child every 30 seconds; this is about three thousand children under the age of five *every day*. About 90 percent of malaria cases occur in sub-Saharan Africa.[21]

■ ■ ■

Unless you are a plant aficionado, you'd miss it. A common weed known as sweet Annie or sweet wormwood (*Artemisia annua*) grows on disturbed roadsides, in fallow farm fields, and in garden beds. It is trodden on by hitchhikers, tires, and snowplow blades, sprayed with every known herbicide. Little noted or respected. An easy weed to overlook.

But it is at present the primary source of the world's most potent antimalarial medicine.

If you pick a leaf and rub it between your fingers, you can't help but notice the smell: pungent, aromatic. And it has a very bitter taste (don't even try it). The Greek for *Artemisia*, *absinthian*, means "undrinkable." The absinthe of France is distilled from a species of this plant, and absinthe, the "green fairy," has its own remarkable history, but that's a story for another time.

As we know, plants are innate chemical factories. And those chemicals have been of interest since civilization began. Why? Because before Walgreens came along, people used plants for medicine, from willow bark (the basis of aspirin) to poppy sap (morphine). Many people still do for that matter. So for malaria, people looked to plants for a possible cure.

A thousand years ago in China, where sweet Annie is referred to as *qinghao*, it was discovered that this plant could kill *Plasmodium falciparum*, the most common malarial parasite. How? The plant produces a unique chemical compound called artemisinin (figure 7.2) with a unique peroxide bridge that scientists believe oxidizes the parasite.[22]

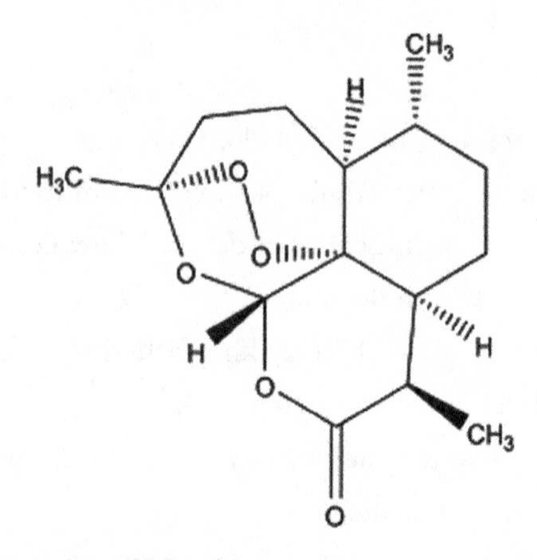

FIGURE 7.2 Artemisinin, a plant-based chemical compound used to treat malaria.

A cure? Not entirely. Artemisinin kills most of the parasites, but they can quickly return. The solution? Combine artemisinin with some slower-acting quinine-based drugs like amodiaquine, mefloquine or sulfadoxine and lumefantrine. The slower-acting partner drugs clear the remaining parasites. This approach is called artemisinin combination therapy, or (ACT). The World Health Organization (WHO) recommends ACT as the first line of defense against *P. falciparum* malaria worldwide.[23] While new strains of *Plasmodium* are emerging that are resistant to artemisinin, no alternative to artemisinin derivatives yet exists, and their efficacy should be maintained.

So how does more CO_2 fit into this picture? We know that it makes plants grow more, but can it also affect plant chemistry? More to the point, can it affect the concentration of artemisinin in sweet Annie?

Let's find out. Our hypothesis is this: Suppose that the increase in atmospheric CO_2 during the twentieth century has altered the concentration of artemisinin in sweet Annie. We'll begin in China, where the medicinal properties of this plant have been recognized for some time. Indeed, many Chinese herbariums (a fancy name for plant collections) have kept pressed samples of sweet Annie for the last one hundred years.

That is a critical factor, because if we have samples from a period of one hundred years, those samples should reflect the increase in atmospheric CO_2 of about 25 percent (from 296 to 370 ppm) that occurred over the course of the twentieth century, with most of that increase occurring since 1970.

Now I turn to an esteemed colleague, Dr. Chunwu Zhu of the Chinese Academy of Sciences, who is the originator of the hypothesis of CO_2 and artemisinin. And a go-getter with the wherewithal to travel around the country to obtain *qinghao* leaf samples from the herbariums shown in the map in figure 7.3.

His hypothesis was pretty simple: If an increase in atmospheric CO_2 increases the concentration of artemisinin in *qinghao* leaves, that increase should mimic the increase in the atmospheric concentration of CO_2 observed during the twentieth century; that is, slow at first, with a greater increase from 1970 onward. Figure 7.4 illustrates what he found.

Interesting. The increase in the concentration of artemisinin appears to mirror the increase in atmospheric CO_2. But maybe some other factor is behind the increase? More fertilizer? Or something else? This was easy enough to check, and Chunwu did so by increasing the concentration of CO_2 in a field study of

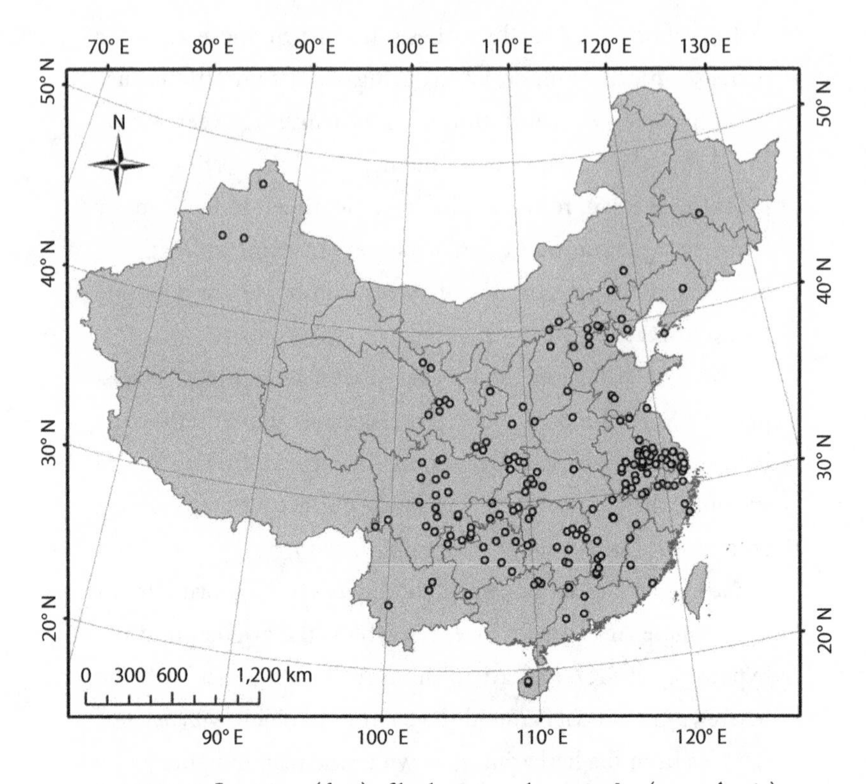

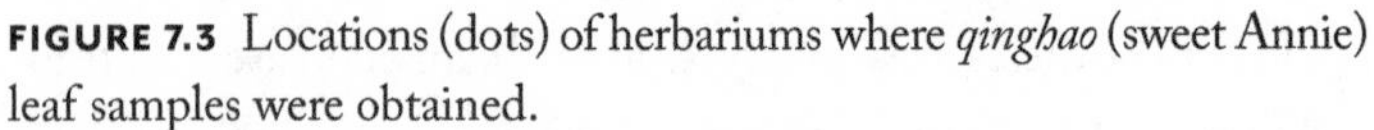
FIGURE 7.3 Locations (dots) of herbariums where *qinghao* (sweet Annie) leaf samples were obtained.

From C. Zhu, Q. Zeng, A. McMichael, K. L. Ebi, K. Ni, A. S. Khan, J. Zhu, et al., "Historical and Experimental Evidence for Enhanced Concentration of Artemesinin, a Global Anti-malarial Treatment, with Recent and Projected Increases in Atmospheric Carbon Dioxide," *Climatic Change* 132 (2015): 295–306.

fertilized *qinghao*. What he found was a higher concentration of artemisinin. Brilliant!

If you want to know more, these data have been published in the journal *Climatic Change*.[24] These are the first data to provide evidence that global changes in artemisinin chemistry could be

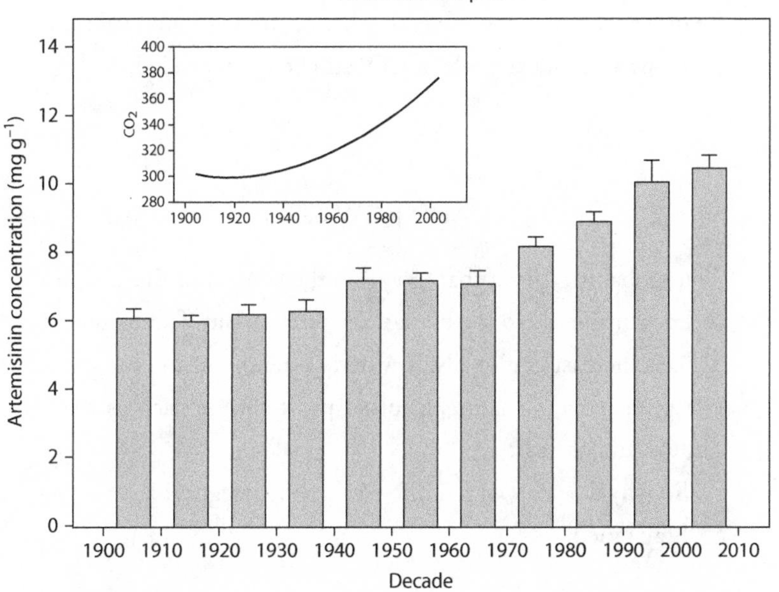

FIGURE 7.4 Carbon dioxide-induced increases in artemesinin, an antimalarial compound found in *qinghao* (sweet Annie), during the twentieth century.

From C. Zhu et al., "Historical and Experimental Evidence."

related to recent increases in atmospheric CO_2. With this information comes the potential of providing a greater quantity of artemisinin from the same area of cultivation. A drug that is essential in controlling malaria.

In this chapter, I have presented evidence of how more CO_2 could benefit humans. The good part. The potential to influence the selection of crop varieties that produce greater yields, to increase marijuana yields, and to increase the concentration of a chemical compound in a plant that can treat malaria. There may be other examples, too, given that the plant kingdom is vast and complex and that the role of CO_2 still being unraveled.

AND NOW . . .

You probably noticed that "good" in the context of this chapter referred primarily to the human perspective. But if atmospheric CO_2 is increasing globally, it would be remiss of me to ignore the perspective of nature, of other plant species such as trees, shrubs, and grasses.

So why doesn't *that* fall into the "good" category?

Stay tuned.

Chapter Eight

CO_2 IS PLANT FOOD

THE BAD

THE REST OF THE STORY

I ran the tractor over the concrete pad and killed the engine. I slid down and headed for the closed-off, air-conditioned shed that was cat-cornered from the tractor barn, pulled out my lunch, and collapsed onto one of the threadbare couches. But not before turning on the radio. It was one minute before twelve.

Fred, one of the permanent drivers (I was a summer hire) came in, heard the radio, and grunted, "You forgot to turn on the PA," and flipped a switch. The heavy baritone voice of Paul Harvey filled the room. Well, our room and anyone within shouting distant of the tractor barn. This happened every day at noon. Everyone on the farm ate lunch and listened to Paul Harvey.

Pre-internet, it was our way of hearing the news. But all of us were transfixed by his "behind-the-scenes, you-are-there" moment, which he deemed, "the rest of the story." It was always interesting—insightful—and made us feel that we knew more than the average Joe.

So now, with apologies to Mr. Harvey, let me bring you the rest of the story when it comes to "CO_2 is plant food."

Yes, CO_2 can make wheat and rice (and marijuana) grow more, and maybe even get sweet Annie to produce more artemisinin, an important antimalarial drug. Good stuff. But let's dive a bit deeper into crops and agriculture. A best-guess estimate is that there are about four hundred thousand plant species on Earth, but only a handful are cultivated, with corn, rice, and wheat supplying the bulk (about 50 percent) of plant-based foods. Roughly 90 percent of the global population is fed by about twenty-five plant species.[1]

There are around eight billion people alive today who depend on agriculture and high crop yields, and anything—anything—that interferes with those yields is deadly serious. Since agriculture began, there have been three main biological threats to yield: insects (e.g., locusts, aphids), pathogens (e.g., bacteria, fungi), and weeds. Guess which causes the most damage? If you guessed weeds, slap a star on your forehead. On average, every crop has about eight to ten "troublesome weeds."[2]

How much damage can weeds cause? One hell of a lot. Overall, the damage is twofold: a reduction in production and a reduction in quality. Weeds can reduce the milling quality of rice, the consistency of wheat flour, and, perhaps of deepest concern, the brewing of beer. Weeds within forage crops like hay can also reduce the protein concentration of the crop and the economic return for ranchers. Got weeds? Less crop.

TABLE 8.1 Global estimates of rice yield loss due to weed interference

Weed species	Yield loss (%)
Red rice	82
Barnyardgrass	70
Bearded sprangletop	36
Amazon sprangletop	35
Broadleaf signal grass	32
Hemp sesbania	19
Spreading dayflower	18
Northern jointvetch	17
Eclipta	10

Source: Roy J. Smith, "Weed Thresholds in Southern US Rice, *Oryza sativa*," *Weed Technology* 2, no. 3 (1988): 232–41.

So if CO_2 can make plants grow, what does it do for weeds? Let's look at this a bit more closely, for weeds and rice (table 8.1).[3] (As a reminder, rice is one of the world's most important crops, being a significant food source for more than two billion people.) What are the worst weeds for rice? (Assuming, for now, no weed control.)

So the worst weed for rice is, uh, rice. Red rice is wild or weedy rice; it's the same genus and species as domesticated or cultivated rice. It's an advantage for a weed to be related to the crop it's infesting. It blends in because it looks just like the crop. Identification is difficult; control is difficult. Not surprisingly, in rice fields around the world, many of the worst weeds look like rice and consequently escape detection. (One can imagine weedy rice rubbing its leaves together and cackling uncontrollably as it sneaks into a rice field.)

Humans, of course, need to stop this. To, ahem, nip it in the bud. Weed management varies tremendously from country to country. Table 8.2 provides some figures on the amount of damage that weedy rice can do without chemical or physical control.

The next obvious question is, As CO_2 keeps going up, who wins? The weed or the crop? Yeah, it's what you think (figure 8.1).

The ratio of seed yield produced from a cultivated rice line is expressed relative to the seed produced from weedy or red rice at two densities and three CO_2 concentrations.[4] Seed production is crucial to yields, but as CO_2 goes up, seed production increases more quickly for weedy rice than for cultivated rice. In simple terms, at a CO_2 concentration of 300 ppm (as was the case in the early twentieth century), a cultivated rice line is superior to a weedy red rice line, but as CO_2 increases, the cultivated rice loses its competitive advantage.

Note something important: It isn't that crops can't respond to CO_2. We know that they can (the "good" remember?). *But so can weeds*—and what we generally observe when crops and weeds are grown together with more CO_2 is that crop yields decline.[5] That doesn't invalidate the "CO2 is plant food" meme, but it provides some context, because in the real world weeds exist. They compete.

And when a resource is increased, weeds can do very, very well. Another example of plant competition—from long ago in a cornfield far, far away (cue the *Star Wars* theme)—a farmer had the idea that since weeds and crops competed for resources,

TABLE 8.2 Rice yield loss due to weed interference across six countries

Country	Yield loss (%)
India[a]	30–90
Philippines[b]	57–61
Malaysia[c]	10–42
Vietnam[d]	15–17
South Korea[e]	5–10
United States[f]	6

[a] S. K. Mukhopadhyay, "Current Scenario and Future Lines of Weed Management in Rice Crop with Particular References to India," in *Proceedings of the 15th APWSS Conference (Tsukuba, Japan, 24–28 July 1995)* (Asian-Pacific Weed Science Society, 1995), 17–27.

[b] P. K. Mukherjee, Anindya Sarkar, and Swapan Kumar Maity, "Critical Period of Crop-Weed Competition in Transplanted and Wet-Seeded Kharif Rice (*Oryza sativa* L.) Under *Terai* Conditions," *Indian Journal of Weed Science* 40, nos. 3 and 4 (2008): 147–52.

[c] Rezaul Karim, Azmi B. Man, and Ismail B. Sahid, "Weed Problems and Their Management in Rice Fields of Malaysia: An Overview," *Weed Biology and Management* 4, no. 4 (2004): 177–86.

[d] Duong Van Chin, "Biology and Management of Barnyardgrass, Red Sprangletop and Weedy Rice," *Weed Biology and Management* 1, no. 1 (2001): 37–41.

[e] Soon-Chul Kim and Woon-Goo Ha, "Direct Seeding and Weed Management in Korea," in *Rice Is Life: Scientific Perspectives for the 21st Century. Proceedings of the World Rice Research Conference Held in Tokyo and Tsukuba, Japan, 4–7 November 2004*, ed. K. Toriyama, K. L. Heong, and B. Hardy (Los Baños, Philippines, and Tsukuba, Japan: International Rice Research Institute and Japan International Research Center for Agricultural Sciences, 2005), 181–84.

[f] Alvaro Durand-Morat, Lawton Lanier Nalley, and Greg Thoma, "The Implications of Red Rice on Food Security," *Global Food Security* 18 (2018): 62–75.

FIGURE 8.1 Red or weedy rice (lighter) is favored more with rising CO_2 than cultivated rice (darker), resulting in greater yield losses. As CO_2 increases, weedy rice outcompetes cultivated rice.
Author's photo.

he would provide so much fertilizer (a resource) that there would be no more competition! Plenty for both the weeds and the crop! So he added ten to twelve times the recommended amount of fertilizer to his cornfield and settled back—no weeding this year!

Came the fall, and his corn yields had changed. They were zero. It was the weeds that were as high as an elephant's eye.[6]

Oops.

Why do weeds win the resource battle? Plants compete for resources and use their diversity—in size, shape, height, and flowers, among other factors—to capture those resources. There is no perfect, ideal plant; if there were, only one kind of plant would exist. So beneficial adaptation to rapid change, as in a sudden increase in global atmospheric CO_2, is likely to be related to diversity: The more genetically diverse a given species or group of species is, the more likely that a winning shape, size, or form, for example, will be selected. The more diverse you are genetically, the more likely you are to win the environmental lottery.

When it comes to crops and weeds and a rapid increase in CO_2, there are two categories: crops, which are purposely designed to be narrow in their genetics (e.g., the Russet Burbank potato), and eight to ten weed species that are highly diverse genetically and in size, shape, and rate of development. They are weeds in part because of their *ability to rapidly adapt to a changing environment.*

SPRAY AND PRAY

OK, OK, OK. Point taken. CO_2 is plant food for *all* plants. But hey, have you never heard of Roundup? Please!

Fair point. So what if weeds in crops get more aggressive as CO_2 rises? Farmers can just spray them with chemicals designed to kill weeds: herbicides. So no worries.

Right?

Well, there are some assumptions. The most obvious is that CO_2 won't, in and of itself, alter the effectiveness of the chemicals being sprayed. Yet there are some sound scientific reasons that challenge that assumption. First, more CO_2 causes leaf pores to close and can alter the thickness, size, and shape of leaves (CO_2 is plant food after all)—all aspects that can affect how much chemical gets taken up by the plant.[7] One possibility is that because herbicides often attack the biochemistry associated with photosynthesis, the stimulation of photosynthesis by CO_2 could conceivably increase the effectiveness of the chemical. On the other hand, the best time to control weeds is when they are small, and if CO_2 stimulates weedy growth, then that time could be limited. On the other, other hand, if CO_2 increases leaf area, then perhaps more of the chemical would be absorbed when it is sprayed, increasing the herbicide's efficacy.

What does the science say? Keeping in mind that science is fluid and new studies are being done, here's what I found: So far, no studies indicate that more CO_2 increases herbicide effectiveness, a couple studies indicate no effect,[8] and several studies suggest that more CO_2 reduces the effectiveness of herbicides on weeds.[9]

Of course, this stirs my inner science nerd: How is this magic trick accomplished? Biochemically? Leaf changes? Something else? So I supposed. My plan this time was to look at whether more CO_2 would benefit a particularly nasty weed: Canada thistle. If you're curious, it isn't from Canada (it's actually from southeastern Europe and Asia Minor). It got its moniker during the Revolutionary War, when John Burgoyne invaded Vermont from Canada as part of the British plan to cut off New England. He was met by an American, John Stark, and the Brits were defeated at the Battle of Bennington (the event is celebrated in Vermont on August 16).

Before I get too far afield (bad pun, sorry), one result of the invasion was a new weed that began emerging in large quantities throughout Vermont: a thistle with large, spiny leaves, capable of rapid clonal growth and infestation. Naturally enough, the native Vermonters assumed this was a deliberate ploy on the part of the British, a type of early biological warfare, and called the weed "Canada thistle." (I suspect it was simply a hanger-on in the hay that Burgoyne used to feed his horses.)

But so prolific was the weed that by 1795, shortly after Vermont became a state, one of the state's first laws was directed to the control of Canada thistle. Time has not diminished the threat of this weed, and it is now considered one of the most destructive invasive weeds in North America.[10]

Back to my supposing. I was interested in comparing Canada thistle with soybean, Canada thistle being an invasive weed and soybean being, well, soybean. To do this, my team and I did a three-year study of Canada thistle (*Cirsium arvense*) planted with soybean and found that soy responded a little, but Canada thistle responded a lot. Even when we sprayed it with the recommended amount of Roundup, when grown with more CO_2, Canada thistle didn't die.

Why? Well, we looked more closely. The Canada thistle plants grown at both CO_2 concentrations looked dead after being sprayed—but the ones at elevated CO_2 came back within a few days. Weird. So we dug a little more—literally—and found that CO_2 stimulated above ground or shoot growth much less than that of root or belowground growth for Canada thistle (figure 8.2).

FIGURE 8.2 Canada thistle sprayed with the recommended dosages of glyphosate (Roundup) as a function of CO_2 concentration. At the higher CO_2 concentration, glyphosate failed to kill the thistle.
Author's photo.

What difference would that make? Well, the increase in the root-to-shoot ratio meant a dilution in the roots of the active ingredient of the herbicide (glyphosate). As the roots survived, they regenerated new shoots asexually. Thus, the herbicide failed to control the Canada thistle.[11]

Is this *the* mechanism that causes a reduction in the effectiveness of chemical weed control as CO_2 rises? Doubtful. Weeds are diverse, and the mechanisms of herbicides equally so. The bottom line is that it is unlikely that there is a one-size-fits-all explanation.

Could we have controlled Canada thistle at the higher CO_2 concentration with more Roundup, or if we sprayed more often? Probably. But that would have had other consequences. Economic for one, as it costs a farmer money every time they spray. Health wise for another, the jury is still out, but glyphosate may be associated with non-Hodgkins lymphoma.[12]

There's more. Leaving aside the economic and health implications for the moment, there is evolution. The more you spray, the more likely you are to select for resistance to the chemical you are using; there's a reason why Lysol says it kills 99.9 percent of germs, not 100 percent.

I can distinctly remember going to a conference in the mid-1990s and being told by a Monsanto rep that "never ever, cross my heart and hope to die, would a weed become resistant to glyphosate. Ever." (Hey, I'm paraphrasing.) Glyphosate was the wonder weed killer of the 90s, the penicillin of its time. So much so that GMO soybean (hint: Roundup Ready soybean) now dominates 95 percent of the global market. So much so that agriculture departments around the country—and the world—stopped teaching integrated pest management and alternative weed management. Just spray glyphosate!

But arrogance and ignorance are expensive. For Monsanto, evolution came with a price tag. The more Roundup was sprayed, the more resistance occurred in the population. (In fairness, I'm sure doctors who prescribed penicillin in the 1960s were equally slow to recognize change.)

Figure 8.3 illustrates the number of weeds that are resistant to glyphosate (Roundup) as of this writing.

I suppose when you're faced with immediate threats to your product, projected what-might-be scenarios for CO_2 and herbicide effectiveness don't seem critical. But perhaps they should.

An important rice variety, Clearfield, has a mutation that makes it resistant to the herbicide imidazolinone. (I can't pronounce it either, so think of it by its brand name: Newpath.) Your rice field is infested with red weedy rice? No problem, just spray it with Newpath; Clearfield will live, and the weedy rice will die.

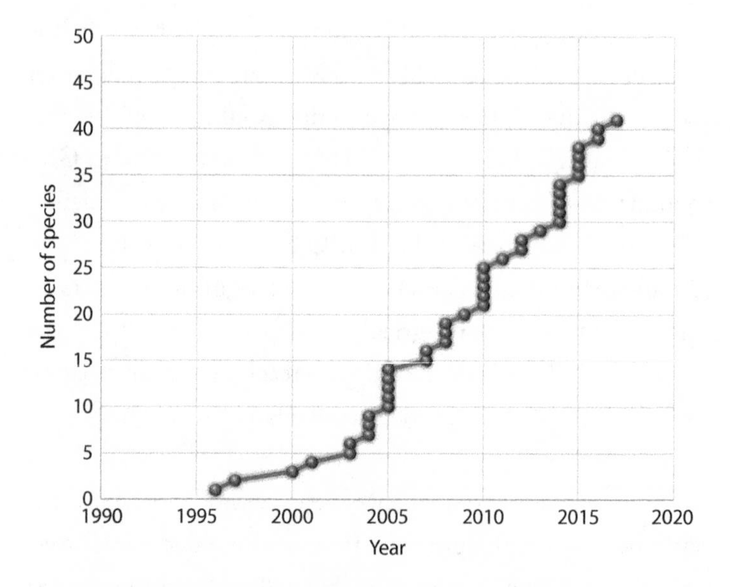

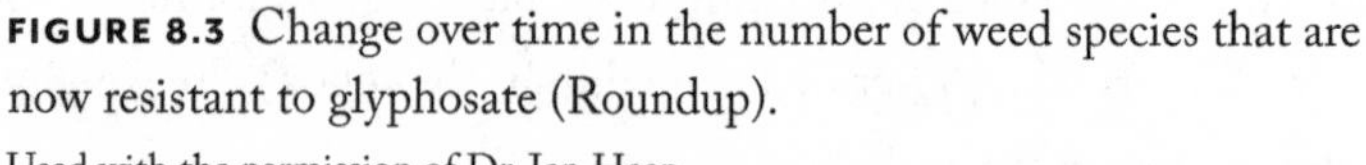
FIGURE 8.3 Change over time in the number of weed species that are now resistant to glyphosate (Roundup).

Used with the permission of Dr. Ian Heap.

Provided, of course, that the gene providing Clearfield with its resistance doesn't get transferred to the weedy red rice. That would be bad.

My team and I demonstrated what happens when cultivated and weedy rice are grown together and more CO_2 is provided: Weedy rice is the clear winner.[13] But that is just part of the story.

CO_2 is plant food and as such can change how quickly a plant develops, including when it flowers. To understand why that is important in the context of weed management, I collaborated with another remarkable USDA scientist, Dr. David Gealy of the Dale Bumpers National Rice Research Center in Stuttgart, Arkansas. (Arkansas leads the nation in rice production.) We did a long-term study growing Clearfield with weedy red rice at a ratio of 7:1 at three different CO_2 concentrations—300, 400, and 600 ppm—concentrations from the early and late twentieth century and one projected for the mid-twenty-first century. Our hypothesis was that increasing CO_2 could increase gene transfer between weedy and cultivated rice, making weedy rice more resistant to the herbicide.

Rice for the most part is self-fertilizing, which is good, no need to depend on a third party like bees or butterflies for pollination. But there is still a small amount of outcrossing: genes from one variety can be transferred via pollen to another variety. Fortunately, at the moment, cultivated and weedy red rice lines don't flower at the same time.

Until you give them more CO_2. Then flowering times become more synchronous. Indeed, we found that the amount of outcrossing, while still small, doubled as CO_2 increased.[14]

The result? A new hybrid: a variety of weedy red rice that is now resistant to imidazolinone-based herbicides—the type of herbicide that is supposed to get rid of weedy red rice. Simply put, more CO_2 may increase herbicide resistance among weedy relatives of commercial crops: *weedy relatives that do the most damage to crop production.*

As we have seen with glyphosate, the number of resistant weeds is rapidly rising (see figure 8.3). What if CO_2 differentially affects herbicide-sensitive and herbicide-resistant weeds?

While more research—*much* more—needs to be done, one study has looked at *Echinochloa colona*, or junglerice, a troublesome weed in rice (that also looks and grows like rice) that can be controlled by the herbicide cyhalofop-butyl.[15] (I know, I know. Dow calls it Clincher; isn't that easier to remember?) Again, the hypothesis was straightforward: Would a future climate with more CO_2 and warmer temperatures favor herbicide-sensitive or herbicide-resistant weeds?

Multiple-resistant and sensitive biotypes of junglerice were compared at warmer temperatures with more CO_2, and guess what? Warmer temperatures and more CO_2 increased the resistance of multiple-resistant junglerice, but not of the sensitive *E. colona*, to cyhalofop-butyl.[16]

I am enough of a realist to know that herbicides, at least in most developed countries, aren't going away. If I farm hundreds of acres, chemical weed control is still the most economically effective means to control weed populations. But I would be a fool to ignore the differential response of weeds and crops to rising CO_2 levels.

PASSING GAS

I am twenty feet in the air alongside a loblolly pine, and I am about to thrust my hands into a large batch of poison ivy growing along the bark.

How did I get so lucky?

It started with a phone call to William Schlesinger, a dean at Duke University, which had received millions of dollars in grant money from the Department of Energy to conduct the largest and longest Free-Air CO_2 Enrichment (FACE) experiment in the country.[17]

Why the funds? To answer two questions. Could trees benefit from more CO_2? And could loblolly pine (and forests in general) absorb enough CO_2 that the global atmospheric CO_2 concentration would stabilize, or maybe, fingers crossed, even start to decline?

Within this magical forest there is a ring—no, not *that* ring—but a metal ring one hundred feet across. And along the ring are forty-foot-tall standpipes, vertical flutes, with holes hissing additional CO_2 to the center of the ring to stimulate pine growth. Adjacent to the rings are little stubs of train tracks for train cars to deliver the six tons of CO_2 necessary, every day, to keep the experiment going.

So if more CO_2 makes pine trees grow more, isn't that a "good" aspect of "CO_2 is plant food"? Maybe. But there is a catch—a big one. The extra CO_2 didn't just make the pine trees grow; it made plants in the forest understory grow *more*.

In fact, the winner in the "who responds more to CO_2" contest, as discovered by Jacqueline Mohan (another brainiac, now an associate professor at the University of Georgia), was . . . drumroll, please . . . poison ivy. And this is where I come in. I had been doing my own research on the response of poison ivy to CO_2 but using indoor chambers. This was fine as far as it went, but outdoor, real-world trials involving CO_2? Nirvana. So I made a phone call, asked a lot of dumb questions, and within a day was tooling down the I-95 corridor.

Now I plunge my hands into the poison ivy (I'm wearing a Tyvek suit and gloves, by the way), and take my samples. Why? Well, the same reason we were interested in Sweet Annie. If CO_2 changes plant chemistry, what does that mean for the ability of poison ivy to induce a rash?

A lot, as it turns out. And not in a good way. More CO_2 increases the toxicity of urushiol, the oil in poison ivy that induces a rash. So CO_2 not only makes more poison ivy but also makes a more virulent form of the chemical that causes rashes, potentially increasing the severity of rashes.[18]

WEED 'EM AND WEEP

Let me state the obvious: Weeds are not found only in farmers' fields; weeds are everywhere. And while poison ivy is one nonagricultural example, there are many, many more.

Perhaps the most egregious of these are plants that are introduced and end up colonizing a habitat so effectively, so completely, that the plant transforms into a weed—a weed that so dominates its new home that biodiversity declines. These weeds, usually nonnative to a given region, have proliferated in recent years—so much so that they are recognized in a separate biological category: invasive or exotic weeds. Millions of acres of productive rangelands, forests, and riparian areas have been overrun by weedy invaders with a subsequent loss of native plants. It has been estimated that more than two hundred million acres of natural habitats, most in the western United States, have already been lost to invasive weeds.[19]

And how does CO_2 alter their biology? Let's get into the weeds (sorry, couldn't resist). We'll start with one of the most prolific invasive weeds in North America: cheatgrass or downy brome (*Bromus tectorum*). Native to Eurasia and the eastern Mediterranean, it appears to have been introduced many times, probably as a contaminant in wheat seed, and was initially found in scattered patches throughout the west. It is a winter annual grass and can grow and set seed quickly following seasonal rains. From its humble beginnings cheatgrass has grown—and grown and grown. Today it occupies about 158,000 thousand square miles in the United States. In simple terms, an area roughly the size of California has become a monoculture of cheatgrass.[20]

That is pretty impressive. So how does cheatgrass succeed? To quote Jeff Goldblum's character in *Jurassic Park*, "Life finds a way." Simply and efficiently. Cheatgrass likes fire; it is the plant version of an arsonist. It grows quickly, sets seed, and dies, and all that dead plant material becomes fuel for future fires. In fact, the U.S. Forest Service has a nickname for cheatgrass: "grassoline."[21] Once cheatgrass comes to your neighborhood, the frequency and intensity of fires spikes. And there are consequences.

You see, the shrub-dominated landscape of the western United States (including sagebrush, piñon, and juniper) can take a brush fire and be fine—*as long as the fire happens only every forty to fifty years*. But when cheatgrass enters the neighborhood, fires can occur every *couple* of years.

The result? Native shrubs disappear, and cheatgrass dominates the landscape (figure 8.4a) from scrub to shining shrub. This is a quintessential example of how an invasive species can significantly alter plant ecosystems with environmental consequences over large geographical areas. There are economic consequences, too: firefighting costs go up, and rangeland values go down.

All that remains is to ask our familiar hypothesis, to suppose. How has cheatgrass responded to recent increases in CO_2 levels? Let's examine intervals of about 50 ppm, from preindustrial times to current conditions. Let's go a step further and look at a range of cheatgrass populations collected at different elevations (1,220, 1,586, and 2,171 meters) in the Sierra Nevada mountains.

Data we obtained from different populations at different elevations revealed the same thing: While there is variation among the populations in how they respond to CO_2, *all* populations tested showed a significant increase in biomass with even small increases in CO_2. Literally more fuel for the fire (figure 8.4b).

It gets worse. As with other plants, it isn't enough to document whether CO_2 makes them grow. We also need to see what happens to their chemistry. In the case of cheatgrass, chemistry is important for a couple reasons. First, it affects the digestibility of cheatgrass for mule deer, their primary herbivore; second,

FIGURE 8.4 Recent increases in the concentration of atmospheric CO_2 may have increased the biomass and flammability of (a) cheatgrass, an invasive plant species associated with (b) fire outbreaks in the western United States.

(a) U.S. Department of Agriculture, public domain. (b) Photo by Joe Bradshaw.

it affects potassium levels, which are inversely correlated with flammability. The worst things more CO_2 could do in terms of fire would be to make cheatgrass less palatable to deer and to decrease the potassium content of cheatgrass. And it does both.[22] CO_2 is plant food.

Lucky us.

Needless to say, cheatgrass isn't the only example of CO_2 making a bad situation worse when it comes to invasive plant species. There are also yellow starthistle, Canada thistle, field bindweed, kudzu, spotted knapweed, leafy spurge, perennial sowthistle, and the list goes on.

So what to do? Spray, right?

Uh-uh. No way. There are eight million acres of kudzu, twenty million acres of yellow starthistle, and one hundred million acres of cheatgrass (remember?).[23] Spray millions of acres with chemicals? Not gonna happen. Covering the western United States with Agent Orange is *not* an option.

But hey, wait. These CO_2 data are for individual plants. What happens in a plant community? Will they still be selected for (outcompete other plant species) as CO_2 increases? Will invasive weeds be the winners, like poison ivy is in pine forests?

Good question, and a harder one to answer. Researchers found that red brome, a relative of cheatgrass grown in a FACE system, was the dominant species in an ecosystem with more CO_2—but only in a wet El Niño year.[24] However, studies in other ecosystems have had different results (table 8.3).

Are more data needed? Always. But the data so far suggest that CO_2 may in fact select for invasive weeds, even within diverse plant communities.

TABLE 8.3 The effect of rising CO_2 on the recruitment of invasive species within plant communities

Species	Community	Favored?
Cherry laurel[a]	Forest understory	Yes
Dalmatian toadflax[b]	Mixed-grass prairie	Yes
Honey mesquite[c]	Texas prairie	Yes
Japanese honeysuckle[d]	Forest understory	Yes
Yellow starthistle[e]	California grassland	Yes

[a]Stephan Hättenschwiler and Christian Körner, "Does Elevated CO_2 Facilitate Naturalization of the Non-indigenous *Prunus laurocerasus* in Swiss Temperate Forests?" *Functional Ecology* 17, no. 6 (2003): 778–85.

[b]Dana M. Blumenthal, Víctor Resco, Jack A. Morgan, David G. Williams, Daniel R. LeCain, Erik M. Hardy, Elise Pendall, and Emma Bladyka, "Invasive Forb Benefits from Water Savings by Native Plants and Carbon Fertilization Under Elevated CO_2 and Warming," *New Phytologist* 200, no. 4 (2013): 1156–65.

[c]H. Wayne Polley, Hyrum B. Johnson, Herman S. Mayeux, Charles R. Tischler, and Daniel A. Brown, "Carbon Dioxide Enrichment Improves Growth, Water Relations and Survival of Droughted Honey Mesquite (*Prosopis glandulosa*) Seedlings," *Tree Physiology* 16, no. 10 (1996): 817–23.

[d]R. Travis Belote, Jake F. Weltzin, and Richard J. Norby, "Response of an Understory Plant Community to Elevated [CO_2] Depends on Differential Responses of Dominant Invasive Species and Is Mediated by Soil Water Availability," *New Phytologist* 161, no. 3 (2004): 827–35.

[e]Jeffrey S. Dukes, Nona R. Chiariello, Scott R. Loarie, and Christopher B. Field, "Strong Response of an Invasive Plant Species (*Centaurea solstitialis* L.) to Global Environmental Changes," *Ecological Applications* 21, no. 6 (2011): 1887–94.

Why? The exact reason is unclear, but we have some ideas. Basically, the most invasive species live in resource-rich environments but are kept in check owing to high pest pressures. But if they get introduced into a new area—with no pests—they can become dominant. So as CO_2 increases—more of a needed resource—such species might benefit more, especially if they are in an environment where their enemies are absent.[25]

SO FAR

Among pests, weeds are acknowledged as the primary contributor to economic loss in crop production,[26] as well as other managed plant systems including rangelands[27] and forests.[28] Weeds are also known to impact natural environments by reducing biodiversity[29] and negatively affecting ecosystem services.[30]

Simply put, weeds affect every human effort to grow a desired plant species, from forests to rice. And while individual crops such as wheat, rice, and soybean can respond to CO_2, when grown in competition with weeds, with few exceptions, crop production will decline. Although loblolly pine can respond to CO_2, when grown in competition with poison ivy, poison ivy is the winner.

Earlier I asked why, given the role of CO_2 in stimulating about 90 to 95 percent of all plant species, the responses of plants such as trees, shrubs, and grasses couldn't fall into the "good" category of "CO_2 is plant food." And the reason, as illustrated here, is that individual plants and communities respond differently to CO_2. There are winners and losers within a plant community as CO_2 rises. So far, poison ivy and cheatgrass are winners (sorry, Grassoline).

But winners may include not only individual species but also particular groups. For example, researchers have found that in tropical forests (where biodiversity flourishes), vines (lianas) respond more to CO_2 than do trees, suggesting both fundamental changes in carbon sequestration by forests and potential reductions in biodiversity if vines become dominant.[31] Why the difference? Well, think about how trees and vines grow. Trees are slow and methodical, laying down layers of support (wood) as they grow taller, shading out competitors as they strive to be first to receive sunlight. Vines are the "bad boys" of the plant world. Taking advantage of the physical support that trees provide, they grow rapidly—they don't have to make wood—and can in effect shade out trees. So when more of a resource is given (CO_2), vines can use it to make more leaves and grow more quickly. Trees? They still need to invest in their infrastructure, wood.

Right now, we do not know if invasive weeds, or vines, will always be the winners in plant communities. But what we *do* know suggests that as the winners and losers become apparent with rising CO_2, diversity will decline. And with a decline in diversity will come a reduction in ecosystem functioning. A diverse ecosystem can support many more animal species (e.g., insects, amphibians, and mammals) than can a monoculture. Hence, the idea that by stimulating plants, CO_2 will result in a "wonderful gift" to nature is absurd.

Given the essential oneness of plants and nature, what else can rising CO_2 do? Let your imagination soar.

Chapter Nine

THE OMG

WHAT WE KNOW

I hope I have provided some evidence that it isn't only the "good" plants that will benefit as atmospheric CO_2 increases but also many of the "bad" plants, the weeds. Weeds are already proliferating and will continue to do so, causing greater and greater damage to both crops and ecosystems in the process. They are not only flourishing but also becoming harder to control, at least by chemical means. This fact alone provides some basis for thinking that "CO_2 is plant food" is not the silver lining of climate change it is made out to be.

Next, I would like to provide some context. To illustrate why the effect of CO_2 on plant biology is beyond a simple good-versus-bad (crops-versus-weeds) dichotomy. As you read at the beginning of this book, plants are the basis of civilization. They twine and proliferate, infiltrate and prosper through all human affairs, from the prosaic, food and medicine, to the aesthetic, music and art, and to the perilous: poison and pesticides.

So come with me for a deeper dive into "CO_2 is plant food" and what it means for you. You and civilization. Human existence—and all of nature.

NUTRITION

We now know that rice responds to CO_2. And that rice is important. It is a staple food for about half the world's population, supplying more than 30 percent of the total protein for 3.4 billion people.[1]

Let's assume that rice will be grown weed free globally in the next few years—not likely, but let's pretend. So as CO_2 goes up, rice yields should increase. But as we have seen, CO_2 can alter plant chemistry, so let's pose another hypothesis, another suppose: Does CO_2 alter rice *quality*—nutrition wise?

Let me pause a moment here to send kudos, praise, prestige to Jann Conroy, an Australian plant physiologist who had the smarts to start asking that question back in the early 1990s for wheat, rice, and other cereals.[2] She found that as CO_2 stimulated growth and yield, it diminished nutritional quality. More specifically, the nitrogen concentrations of foliage, roots, and grain were consistently lower in plants grown at elevated CO_2. Thank you, Jann.

Other scientists were skeptical. I can recall my boss saying, "Well, that's because she didn't have sufficient nitrogen in the soil to begin with." Except she did. The decline in nitrogen with more CO_2 occurred no matter the availability of nitrogen in the soil. Others (including me) repeated her work with rice at the International Rice Research Institute. The results were the same. More CO_2? Less nitrogen.

So what? Who gives a flying fig if nitrogen is reduced?

Nitrogen is a proxy—a value—that reflects something more important: protein concentration. Many people globally get their protein from plants. And the plant protein market is growing—just ask the Beyond Meat entrepreneurs. So if the major cereals—specifically wheat, rice, and corn—have less protein as CO_2 increases, what then?

There's more. There's always more. Suppose (!) additional CO_2 means a change in mineral composition? Some minerals, like iron (Fe) and zinc (Zn), are indispensable for human health. Plants supply what is known as non-haem iron and are the source of about 60 to 70 percent of the iron you need. Iron plays an essential function in animals, for example by helping red blood cells carry oxygen. Not enough iron? You're likely to feel tired, listless. Low iron can also lead to mental impairment and decreased immune system functioning. Iron is also recognized as essential during pregnancy.[3] And zinc? Well, zinc is present in all cells in the human body. It is necessary for proper immune system function, as well as cell division, cell growth, wound healing, and carbohydrate metabolism. In addition, zinc is needed for your senses of smell and taste. During pregnancy, infancy, and childhood, the body needs zinc to grow and develop properly. Zinc also enhances the action of insulin.[4] (But hey, other than that . . .)

So how does more CO_2 affect iron and zinc? Figure 9.1 provides some data for eighteen rice lines from China and Japan,

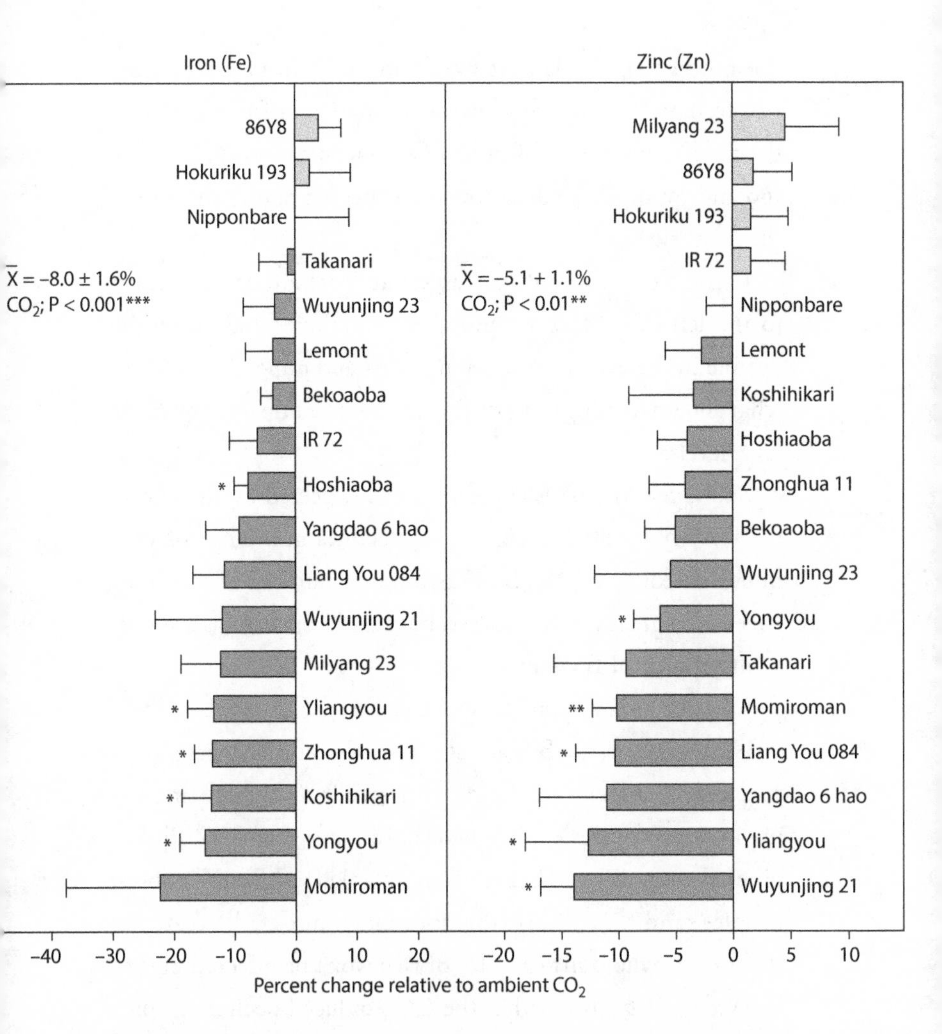

FIGURE 9.1 Changes in the concentration of iron and zinc at projected increases or decreases (plus standard deviation) in concentrations of CO_2 predicted by the end of the century for eighteen genetically different rice lines (varietal names are given). Lines to the left of center indicate a CO_2-induced decline in iron or zinc; lines to the right of center represent an increase.

Figure by the author.

the product of the largest evaluation of iron and zinc from rice to date.[5] The figure shows a comparison of rice grown in the field at current and future CO_2 concentrations of between 569 and 590 ppm, a concentration expected before the end of the century.

Hmm, no significant increases, but there are declines (lines to the left of center) in a number of varieties and an overall decline in the average. We see proteins and minerals declining, what about vitamins from rice? Can those be impacted by rising CO_2 levels?

Sadly, yes. The CO_2 concentrations expected by the end of the century resulted in significant declines in a range of vitamins, including B1, B2, B5, and B9, in a number of rice lines. Rice isn't a great vitamin source, but it is recognized as a significant supplier of B vitamins.

OK, but not everyone eats rice, right?

Not everyone, no. But about six hundred million people get 50 percent or more of their calories *daily* from rice, including in the countries of Bangladesh, Indonesia, and the Philippines, among others. Rice is often a staple food for the poorest people—and they will be the ones most impacted.[6] Yeah, but those of us who don't eat a lot of rice won't be affected, right?

Wrong. It turns out that the CO_2-induced declines in protein, iron, and zinc also occur in other major crops like wheat and potato. Groundbreaking analyses by Irakli Loladze (one of the most fascinating and brilliant people I have had the good fortune to work with)[7] and Samuel Myers at Harvard (Harvard!),[8] as well as a number of other prominent scientists,[9] have made incredible strides in documenting just how extensive the negative impact of more CO_2 may be on plant-based nutrition. And while rice is not widely consumed in the United States, the vast majority of us eat bread (a wheat product)—about a pound a week.

Although CO_2 is reducing the protein content of wheat and rice, there is some good news. It turns out that a separate class of plants doesn't show a decline in protein with more CO_2: legumes. Plants like soybean and peanut fix their own nitrogen and in doing so show no effect of CO_2 on protein concentration.[10]

Nice, but it's hard to imagine that the world will stop eating bread and rice and focus on soy and peanuts. Although it would make for interesting airline snacks . . . Oh, wait.

This finding seems, well, isolated, forlorn. If CO_2 is really reducing nutritional quality, why isn't anyone talking about it? Surely it must be a mistake? Let's dig a little deeper.

One of my favorite movies is *Contact*, based on a book by Carl Sagan, the late astronomer from Cornell and a personal hero whose books inspired me to pursue a scientific career. (I even wrote to him once—and he wrote back! It was *awesome.*)

In the movie, which imagines the first interstellar contact with the human race, Jodie Foster plays the hero, Dr. Ellie Arroway, a SETI (search for extraterrestrial intelligence) scientist who has waited her entire life for an extraterrestrial "message." And waited and waited . . .

And when she first hears a signal after many, many years—a signal that could be her dream, her hopes fulfilled—her response is not "Hooray!" or "Wow!" Instead, she turns to the others in her group and asks them, in essence, to "prove me wrong."

This may seem strange, you wait for years for an event and when it comes, you want to be proven wrong. But her reaction, as Sagan rightly notes, is the essence of doing good science. Science doesn't exist to confirm your own beliefs or prejudices. So if I get an outcome that confirms my bias, I want some alternatives. What other explanations could there be?

So when it comes to crops, nutrition, and climate change, let me provide alternative hypotheses; let me try and prove "CO_2 reduces nutrition" wrong. (I know, I know. For some folks, it doesn't matter. They think scientists are only doing it for the money. Perhaps some are, but my experience has shown that most are not. They are interested in understanding how the world around them functions. For myself, finding out how something works, the magic behind the scenes, is my driving force. I also know that when I publish something, other scientists will not hesitate to call me out if I make assumptions that reflect my own bias—just as I wouldn't hesitate to point out another scientist's bias. It's a two-way street.)

In agricultural science, crop breeders have worked hard to improve the yields of every plant species consumed by humans. To my knowledge, none of the many varieties of produce—from apples to zucchini—that you buy in the supermarket exists "as is" in nature. They have been worked on, tweaked, and massaged to provide as much yield in as little space as possible. (Anyone who says they don't eat genetically engineered food is, well, lying.)

But such breeding and selection efforts come at a cost. Studies have shown mean declines of 5 to 40 percent in the concentration of some minerals as well as declines in the concentration of vitamins and protein in response to traditional twentieth-century breeding.[11] Indeed, if you plant low- and high-yielding varieties of broccoli and cereals side by side, you will find consistently negative correlations between yield and the concentrations of minerals and protein—a genetic dilution effect.

So there is a "time stamp" when it comes to reduced nutrition; perhaps the decline we are currently observing is related to the intensification of yields. But keep in mind that CO_2 has *already* jumped by 25 percent since the 1970s. The same crop varieties tested at increased CO_2 concentrations that have already occurred should be relatively insensitive to CO_2. This is one reason, as I mentioned earlier, we are testing lines from the 1950s and 60s against recent CO_2 changes. Figure 9.2 provides some preliminary data for changes in the concentration of protein in rice owing to changes in CO_2 concentration that have already occurred.

Again, note the significant decline in protein. Each rice variety was given fertilizer on a daily basis, so it seems unlikely that the decline is related to nutritional status and not CO_2. And the decline appears in both low- and high-yielding rice lines.

There needs to be more work in this area to document how crop nutrition has responded to recent CO_2 increases. But it can be difficult to do so. From a methodological viewpoint, it is very hard to remove CO_2 from the atmosphere; for this study, the rice was planted in specially built growth chambers that allowed us to "dial in" earlier CO_2 concentrations.

Let's do another test. Beyond crops. Humans have been messing about with cereals like rice and wheat for several thousand years. So maybe the decline in nutritional quality I'm

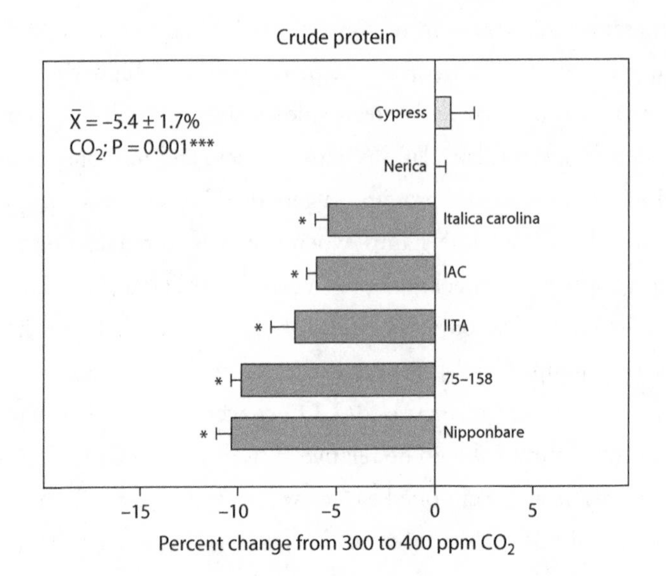

FIGURE 9.2 Changes in the concentration of crude protein in seven rice lines relative to changes in CO_2 concentration during the twentieth century. Five of the seven lines tested showed a significant decline in protein.
Unpublished data from the author.

seeing in rice is some inherent effect of breeding. Is a decline in quality also occurring in native plants? How about I look at a plant that hasn't been selected and bred over countless generations? How about goldenrod?

Goldenrod (figure 9.3) is a fall perennial with bright yellow flowers, a harbinger of winter. It has not been selected for, bred for, or genetically manipulated by humans. It is a roadside weed. (For many years it had a reputation as a fall allergen, but the reputation was unfounded; it happens to flower at the same time as ragweed.) So let's employ another hypothesis: Have

FIGURE 9.3 Canadian goldenrod (*Solidago canadensis*).
Wikimedia Commons.

recent increases in atmospheric CO_2 reduced the protein content of goldenrod pollen?

That's hard to test experimentally. As mentioned, it's easy to add CO_2 but hard as hell to remove it. So let's try a different approach.

Let's go to the Smithsonian National Museum of Natural History! It's one of the highlights of anyone's visit to Washington, DC, with its mammoth mammoths, wondrous whales, and the Hope Diamond.

But way in the back, far from the crowds, next to the Ark of the Covenant (OK, I made that last bit up—too many viewings of *Indiana Jones*), are gray metal cabinets full of goldenrod samples from the 1840s, 1880s, 1920s, and so on.

Why is that important? Because those samples reflect the rise in CO_2 that occurred during the Industrial Revolution. If more CO_2 does reduce nutrition, then goldenrod protein should decline in response. And remember that CO_2 has risen by 25 percent just since 1975.

In the Smithsonian samples, we did indeed find that the protein concentration had declined, at least as indicated by nitrogen proxy—by about 30 percent from the mid-nineteenth century to today.[12] But what about future CO_2 increases?

Here we turn to another excellent USDA scientist: Wayne Polley in Temple, Texas. He and others came up with the Conestoga wagon plan. Sounds weird, but it's nothing short of brilliant.

Imagine, a linear chain of clear plastic chambers ("Conestoga wagons") stretched across the Texas prairie (figure 9.4). Each chamber is linked to the next by a small tunnel. A future (higher) concentration of CO_2 is introduced at one end, then slowly makes its way to the other end of the wagon chain. As it does so, CO_2 is removed by the plants inside the chambers via photosynthesis (CO_2 is plant food!), so that by the time air gets

FIGURE 9.4 Wayne Polley's "Conestoga wagons" in Waco, Texas. At one end are future CO_2 conditions; at the other are past CO_2 conditions.
Photo by Wayne Polley.

to the last wagon, the CO_2 concentration is the same as it was 150 years ago.

The result is a CO_2 gradient that mimics past and future concentrations, fortunately one that allows goldenrod to keep growing. And while the absolute values of protein were different from those in the Smithsonian samples, the conclusion was the same: Recent and future CO_2 increases reduce the protein content of goldenrod.[13] So there may be a breeding consequence associated with nutrition; but so far (and we welcome new data) there appears to be a CO_2 consequence as well.

Pfft. Who cares about goldenrod? Sure, it's kinda pretty, but so what if pollen protein goes down? People don't eat goldenrod.

In our research on goldenrod, why did we choose to look at pollen?

Because if rising CO_2 is affecting human nutrition, it is also likely to affect the nutrition of other species within the food chain. And one of the most critical species in that food chain is pollinators. Among pollinators, almost everyone is familiar with bees, specifically domestic bees (of the *Apis* genus). They are a crucial aspect of modern agriculture. Without them, human food would be a generic and colorless blend of cereals, but with them, we have a rainbow of fruits and vegetables that pop on our plates.

If CO_2 is reducing the protein content of pollen, there is reason for concern. A little background. Yes, bees eat nectar; it's their primary carb source, and they are damn good at locating and identifying these sources—hence the "waggle" dance bees use to communicate with one another, "Hey, turn left at the 7-Eleven, go ten yards, and then you'll see a big sunflower." But the other source of bee food is pollen. If nectar provides carbs, pollen provides protein, and protein is essential for bee health. It is particularly important during the fall because pollen collection and storage are necessary for bees (and their queen) to survive the winter.

And bees aren't as well equipped to distinguish pollen. One beekeeper told me that nearby construction was producing concrete particles about the same size and shape as pollen—and bees were bringing the concrete bits back to the hive. Not good. Several publications have indicated that there is no "waggle dance" for pollen as there is for nectar.[14]

So what do the experiments indicate so far? We have historical and experimental evidence that rising CO_2 can reduce the protein content of pollen. The Smithsonian goldenrod collection indicates about a 30 percent drop in pollen protein since the 1840s; the Texas prairie CO_2 Conestoga wagon experiment confirmed that drop and suggested that it is likely to continue in the near term as CO_2 continues to go up.

If more CO_2 reduces pollen protein, there are consequences. Bees are already experiencing environmental stress from Varroa mites, neonicotinoids (a new class of insecticides similar in chemistry to nicotine), and agricultural monocultures. The implications of rising CO_2 levels for bee health are still widely unknown but need to be understood. Of course, it isn't just humans and bees who will be impacted but likely all animals, from grasshoppers to cows.[15]

As always, I want to know the mechanism behind the observation, the patter behind the trick. So let's dig a bit deeper into exactly how CO_2 reduces the nutritional quality of food. Dilution perhaps, by stimulating plant growth? Or by causing changes in root-to-shoot ratio? Or something else? We're still looking, but in doing so, we have discovered some interesting patterns. One of the most interesting was found in the study on the impact of more CO_2 on the nutritional quality of rice (figure 9.5).

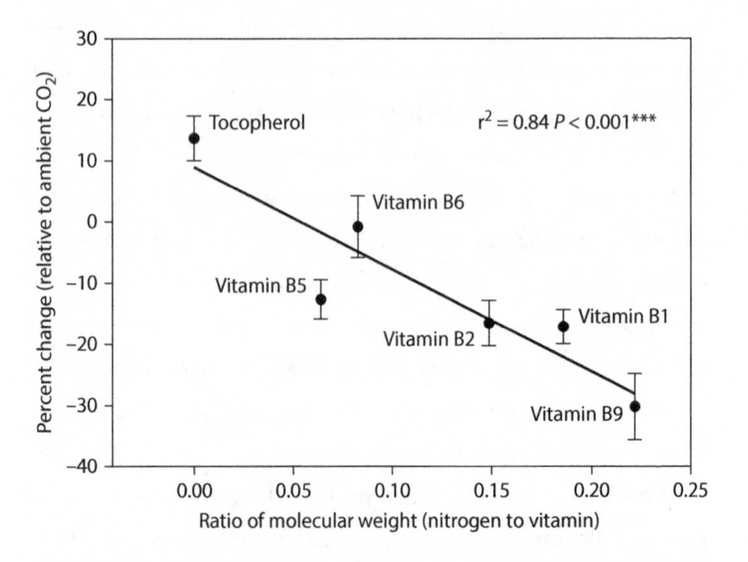

FIGURE 9.5 Percent change in the concentration of vitamins in rice at CO_2 levels expected before the end of the century relative to the amount of nitrogen contained in the vitamin. As CO_2 increases, nitrogen concentration increases and vitamin concentration decreases. Figure by the author.

This pattern is interesting because in addition to the B vitamins, we also looked at tocopherol, or vitamin E, and found that its concentration in rice seed went *up* with more CO_2.

Why does the concentration increase for some vitamins but decrease for others? It's a head-scratcher, but our working hypothesis is reflected in figure 9.5. Basically, there is a negative correlation between the amount of nitrogen contained within the compound and its response to more CO_2. If there's no nitrogen, more CO_2 might increase the concentration of a compound. (Figure 9.5 shows the result for tocopherol, but the same is found with artemisinin, the antimalarial compound discussed in chapter 7. It, too, has no nitrogen, and its concentration also appears to have risen as CO_2 increased during the twentieth century). If there's lots of nitrogen, the concentration of a compound may decline. In basic terms, plants appear to be adapting to changes in the atmosphere. More carbon but less nitrogen? Make more carbon-rich compounds and fewer nitrogen-requiring compounds. Plants make the best use of the resources available to them.

And protein involves a lot of nitrogen, via amino acids that make up protein. So more CO_2 may mean reductions in protein or reductions in compounds that have a lot of nitrogen, from rice to goldenrod pollen.

Is this the future of food? Unclear. We need to do more to determine how common such a response is, but for now, it's a working hypothesis with a number of confirmations. But! If the hypothesis holds, then the nutritional quality of plants—*and all aspects of the global food chain*—will be impacted, and not in a good way.

How not good? Reams have been written about human nutrition (e.g., "an apple a day keeps the doctor away," "you are what you eat"). These sayings highlight the link between proper nutrition and good health; from COVID-19 to heart disease and from cancer to diabetes, we recognize that diet and nutrition are key tools in our medical arsenal.

When poor nutrition is mentioned in America, you might imagine "poor people in other countries," images of undernourished children in Ethiopia or poverty-stricken villages in Burundi. But that doesn't do hunger justice. In July 2020, the

American Journal of Clinical Nutrition highlighted that diet-related illnesses in the United States are rising—quickly—killing more than half a million people each year.[16] And while this problem is related to the COVID-19 pandemic, as well as to unemployment and poverty, kick your imagination into gear, and think about what will happen to millions, even billions, of people as CO_2 continues to rise and the nutritional integrity of our food declines.

ACHOO

If you're like me, springtime is intense. In part because a reawakening of dormant plant life is awe-inspiring, from the yellow of forsythia to the rainbows of primrose.

And the pollen from oak. And birch. And willow. And elm . . .

Are oaks majestic? No question. But they practice what many tree species do in the spring: sex on the cheap. Sex on the cheap is simple; rather than investing in fancy flowers with bright colors to attract pollinators who move the pollen (male gametes, or sperm cells) to the pistil (the female reproductive part of a flower, including the ovary, stigma, and style), some trees just make pollen and rely on wind to get it to their female counterparts. But because wind is not a reliable pollinator, some trees like Oak must make lots and lots and lots of pollen. A huge springtime orgasm. And people with allergies suffer accordingly. Me included—well, me and about twenty-five million others in the United States.[17]

So how does more CO_2 fit into this picture?

Let's switch from trees to an easier (smaller) "sex on the cheap" plant: common ragweed. A single ragweed plant can produce more than a billion pollen grains. In fact, ragweed is among the top pollen offenders, even relative to some of the worst tree species.[18]

Let's delve into our bag of supposes and ask, How does rising CO_2 affect the pollen production of ragweed? Frances Caulfield and I asked this question back in the late 1990s.[19] We grew ragweed at preindustrial, current, and projected CO_2 levels (270, 370, and 600 ppm, respectively). We found that ragweed responded well to both recent and projected increases in CO_2—doubling in size from 270 to 370 ppm and again from 370 to 600 ppm.

And as size increased, so did pollen production (figure 9.6).

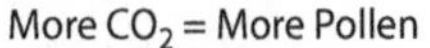

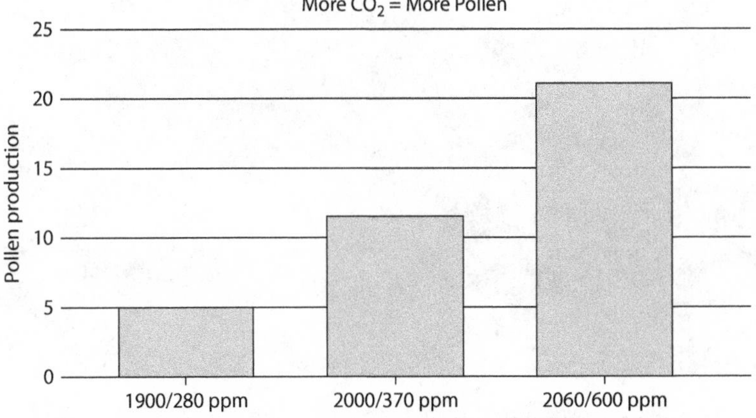

FIGURE 9.6 Average pollen production from ragweed in response to ongoing increases in atmospheric CO_2.

Used with the permission of Climate Central.

There's more. Ben Singer, an inquisitive med student (the best kind), asked the next logical question: Does CO_2 affect the concentration of the allergen on the surface of the pollen, the allergen that triggers the immune system? The short answer is that it does: It goes up as CO_2 goes up.[20] There's still more. Jae-Won Oh and a contingent of innovative South Korean scientists have repeated the ragweed work but using sawtooth oak trees, another major pollen source (figure 9.7). Consistent with the results of the other studies, these researchers found that more CO_2 results in an increase in pollen production by oak

FIGURE 9.7 Sawtooth oak trees growing at different CO_2 concentrations. Increases in CO_2 cause increases in the concentration of the allergen that causes hay fever (allergic rhinitis).
Photo courtesy of Dr. Jae-Won Oh.

TABLE 9.1 Change in the concentration of the allergen that causes hay fever (allergic rhinitis) in sawtooth oak trees as CO_2 increases

CO_2 concentration (ppm)	400 ppm (ambient)	550 ppm (ambient × 1.4)	700 ppm (ambient × 1.8)
Concentration of *Quercus* a 1 (μg/mL)	229.1	287.9 (1.26 times)	353.8 (1.54 times)

Source: Kyu Rang Kim, Jae-Won Oh, Su-Young Woo, Yun Am Seo, Young-Jin Choi, Hyun Seok Kim, Wi Young Lee, and Baek-Jo Kim, "Does the Increase in Ambient CO_2 Concentration Elevate Allergy Risks Posed by Oak Pollen?," *International Journal of Biometeorology* 62, no. 9 (2018): 1587–94.

trees (table 9.1) and an increase in the allergen on oak pollen (*Quercus* a 1) that stimulates people's allergic reactions.[21]

The take-home message is simple: more CO_2 means bigger ragweed and oak plants—and more pollen, more *allergenic* pollen. For some, this is not a big deal—a few more sneezes, the need to remember to bring some Kleenex.

But for others, it's misery. Sinus pressure; watery, itchy eyes; scratchy throat. And for a select few (yours truly among them), asthma, a blocking of the airways and the inability to breathe; sinusitis, with infections of the ears or lungs; or anaphylaxis—severe allergies can increase the risk of your body seizing up and shutting down. (I do not go anywhere without an inhaler. Not being able to breathe is scary as hell.)

Does more pollen result in death? Not directly. (Although if you have a prolonged asthma attack, it feels like you're dying.) But Dutch researchers looked at other aspects of high-pollen days (owing to *Poaceae*, an allergenic grass species found throughout the Netherlands) and found that days with the highest pollen counts were associated with an increase of about 6 percent in death from heart disease, a 15 percent increase in death from chronic obstructive pulmonary disease, and a 17 percent increase in death from pneumonia.[22]

So I went online, and with the help of László Makra, a brilliant Hungarian professor, combed through every database we could find. And we found that, indeed, pollen counts have risen and are continuing to rise around the globe, or at least in the northern hemisphere.[23] How much of this increase is related to CO_2 and how much is related to rising temperatures is still being looked at, but overall, there is a clear link. More CO_2, more pollen.

FOOD ALLERGIES

As mentioned, legumes, plants that fix their own nitrogen, are less likely to be impacted by rising CO_2. But here, too, there are questions: Are certain proteins more sensitive to rising CO_2 than others? If so, what are the consequences for human health?

Data presented at the American College of Allergy, Asthma and Immunology Annual Scientific Meeting suggest that peanut allergy in children has increased 21 percent since 2010 and that nearly 2.5 percent of U.S. children may have an allergy to peanuts.[24] The threat of such allergies has resulted in the removal of peanuts as snacks on airplanes and elementary schools.

Do peanuts respond to CO_2? Yes, just like wheat, rice, soybean, cowpea, yada, yada. Growth and seed yield show a positive response. As other studies have found, there appears to be variation among peanut varieties in their response to more CO_2.[25] For example, in a field study my colleagues and I did, we observed that more CO_2 (about 250 ppm more) resulted in a greater stimulation of seed yield in Virginia Jumbo relative to Georgia Green.[26] So here again is a potential opportunity to select for CO_2 responsiveness, this time in peanut lines. But there is a catch. (Big surprise, huh?)

The stimulation of Virginia Jumbo was simultaneous with an increase in the concentration of a protein called Ara h1. This protein, unfortunately, is the allergen that causes peanut allergies. Increases in CO_2 may open an entire Pandora's box of food allergies. But at present, it is a box whose contents are still murky. Obviously, if more CO_2 can alter the concentration of plant proteins associated with food allergies, that will have a significant impact on associated allergic diseases and pose a potential threat to human health. How much of a threat remains to be determined. But no one wants to die from anaphylactic shock.

A BITE OF THE APPLE

We think of plants as sedate and passive. Nothing could be further from the truth. If you study plants, you will find that they are active competitors, climbing skyward for light while out-shading those nearby, using their root systems to outcompete other plants for scarce nutrients, or adding poison to the soil to reduce the growth of their neighbors. And as for animal interactions? Fascinating. Enticing some animals with rich colors and food rewards (nectar) to spread their seeds and repelling others by making their leaves toxic for consumption. Including us.

Humans have learned over millennia what is and what isn't edible. Apples, sure. Apple seeds? Well, they contain a plant compound known as amygdalin. If the seeds are damaged, chewed, or digested, amygdalin degrades into hydrogen cyanide. It won't kill you if you eat one seed, but it can make you very sick if you eat too many. (There's a reason we throw apple cores at our enemies.) Rhubarb, sure. Rhubarb stalks are tart and are used in many culinary traditions. Rhubarb leaves? Only if you want to die—they contain oxalic acid. Red kidney beans contain toxins called lectins that kill the cells in your stomach. Yuck. Want to eat them? They must be boiled for at least five hours to remove the toxin.

Nightshades? One of the most infamous is *Solanum linnaeanum*, "the apple of Sodom." As the name suggests, it's not something you want to eat. But aren't tomatoes and potatoes also in the nightshade family? Sure, and you can eat them. Well, just don't eat any of their leaves, stems, or sprouts. Or any potatoes with a greenish tint. How scared were people of nightshades? Well, when they arrived in Europe from the New World, people would eat tomatoes on street corners, and observers were convinced they would consequently turn into werewolves. The scientific name for tomato? *Lycopersicon*—literally "wolf" (*lyco*) "peach" (*persicon*).

There are thousands of known poisonous plants. Given that CO_2 can alter plant chemistry, how might more CO_2 affect the toxicology of these species? I wish I could clap my hands and respond, "So glad you asked that question; we know a whole lot!" But of course we don't. But here's a little insight from one of the few studies that has tried to address this issue.

Roslyn (Ros) Gleadow is a curious scientist at Monash University in Australia, and she was interested in cassava, a tuber food crop that is a staple for about 750 million people worldwide. (After rice and corn, cassava is the largest source of carbs in the tropics.) It is a drought-tolerant crop, one that has piqued interest as a stable food source if drought frequencies increase globally.

But cassava has sweet and bitter lines. The bitter lines can deter pests and animals and are bitter for a reason: the tubers contain residual cyanide. As such, they must be properly prepared for eating. If not, cyanide can build up in the human body with side effects including goiter, ataxia, paralysis, and death. One of the most common diseases caused by insufficiently prepared cassava is konzo, a neurological disease characterized by a rapid onset of irreversible spastic paralysis similar to cerebral palsy and multiple sclerosis.[27] In sub-Saharan Africa, especially in Uganda, Tanzania, and the Democratic Republic of the Congo, thousands of people may have experienced cyanide poisoning and konzo from cassava, but sadly, reliable data are unavailable.

Ros was interested in a simple hypothesis: Does more CO_2 make more cassava and/or cause changes in cassava chemistry or nutritional quality? To test this hypothesis, she grew cassava at current and two future CO_2 concentrations (550 ppm and 710 ppm, levels likely to occur near and after the end of the current century). In contrast to what was observed for cereals like wheat and rice, cassava tubers declined in yield per plant and experienced no change in cyanogenic glycoside concentration. (A cyanogen is the substance that induces cyanide production.) However, the concentration of cyanogenic glycosides increased in the leaves, which are also sometimes eaten.[28]

Ros's work leads to a more fundamental question: Can more CO_2 affect the toxicity of the food we eat? Around 10 percent of all plants and 60 percent of crop species produce cyanogenic glycosides.[29] Do they respond like cassava? I wish we knew.

What about nonhuman consequences? Plants are a source of food for all animals, including insects. For hundreds of millions of years, there has been a constant evolutionary struggle between plants and insects involving some pretty sophisticated chemistry, including insecticides plants create to keep from being eaten alive.

Japanese beetles (*Popillia japonica*) are vile. Utterly unredeemable, shiny scarabs that multiply quickly and infest any patch of green they find. A destroyer of fruit, gardens, and field crops, with a total host range of more than three hundred plant species. Currently the Japanese beetle is the most widespread pest of turfgrass and costs the United States turf and ornamental industry approximately $450 million each year in management alone.[30]

When insects, including Japanese beetles, eat their leaves, soybeans and some other plants produce a hormone called jasmonic acid, which initiates a chemical reaction. The result? The production of a protease inhibitor that, when ingested, inhibits the ability of insects to digest leaf material.

Unless you provide more CO_2. Ten years ago in a field in Illinois, soybean grown at a CO_2 concentration of about 550 ppm produced less jasmonic acid—and experienced much more damage from Japanese beetles.[31] Of course, jasmonic acid isn't the only toxin in a plant's chemical armory. Look at tobacco. What, you thought the tobacco plant made nicotine specifically for R. J. Reynolds? For centuries, gardeners have used homemade mixtures of tobacco and water as a natural pesticide to kill insect pests.

And with more CO_2? My colleagues and I looked at this with recent and projected CO_2 changes. And, as with jasmonic acid, more CO_2 resulted in a significant decline in the nicotine concentration of tobacco.[32] Not so good if you're a smoker. Great if you're an insect with a gastric fondness for tobacco leaves.

This is just scratching the surface of plant toxins. Caffeine, in addition to being the world's most popular psychoactive substance, is also a natural insecticide, one that can destroy insect digestive systems. (Don't tell Starbucks.) And more CO_2? Well, for *Camellia sinensis*, used worldwide for tea, more CO_2 (800 ppm) results in an increase in the concentration of carbohydrates in the leaves, as well as an increase in the concentration of polyphenols, free amino acids, and theanine. But caffeine concentration decreases, by about 25 percent.[33] Recent work I published with Fernando Vega, a coffee expert for the USDA, and others also showed a decline in the concentration of caffeine in coffee but curiously only for one species we looked at.[34]

We are, at the moment, just skimming the surface of the interplay between rising CO_2 and plant chemistry and the implications of that interplay for insect–plant interactions. But those interactions are damn important.

TAKE YOUR MEDICINE

Poison is not as far from medicine as one might think. A change in dosage or a slight tweak in chemistry may make the difference between life and death.

Something early civilizations discovered. The ancient Sumerians referred to the poppy plant as the "plant of joy" (*hul gil*).[35] I'm pretty sure they had some idea what it did. And how powerful it was. Powerful enough that knowledge of how to grow poppy has been in the public domain since 3400 BCE.

Of course, the Sumerians have been replaced by GlaxoSmithKline (whose Australian opiates business was bought out in 2015 by Sun Pharmaceutical) and Sumer by Tasmania.[36] At present, Tasmanian commercial poppy fields account for about a quarter of the world's morphine and codeine, as well as new analgesics derived from poppy, like thebaine, which is used to make OxyContin.

Since civilization began, plant chemistry has been recognized as a primary source of human medicine. Table 9.2 provides a partial list of plant-based medicines.

You might think, logically, that drug companies don't rely on plant-based sources. Having a supply chain that involves farming a medicinal plant is messy and expensive; it would be better to synthesize drugs (well, after they've been discovered in plants). And they do. Most of the time. However, at present, about 15 percent of pharmaceuticals in developed countries are still plant based (e.g., morphine from poppy). And, interestingly, that number jumps to about 80 percent in developing countries.[37] It turns out that not every street corner is home to a Rite Aid. Who knew?

TABLE 9.2 Plant-based medicines and their clinical uses

Drug	Action/clinical use	Species
Acetyldigoxin	Cardiotonic	*Digitalis lanata*
Allyl	Rubefacient	*Brassica nigra*
Atropine	Anticholinergic	*Atropa belladonna*
Berberine	Treats bacillary dysentery	*Berberis vulgaris*
Codeine	Analgesic, antitussive	*Papaver somniferum*
Danthron	Laxative	*Cassia* spp.
Ephedrine	Antihistamine	*Ephedra sinica* (Mormon tea)
Galantamine	Cholinesterase inhibitor	*Lycoris squamigera*
Kawain	Tranquilizer	*Piper methysticum* (kava)
L-DOPA	Anti-Parkinsonism agent	*Mucuna* spp.
Lapachol	Anticancer, antitumor	*Tabebuia* spp.
Ouabain	Cardiotonic	*Strophanthus gratus*
Quinine	Antimalarial	*Cinchona ledgeriana*
Salicin	Analgesic	*Salix alba*
Taxol	Antitumor	*Podophyllum peltatum*
Vasicine	Cerebral stimulant	*Vinca minor*
Vincristine	Antileukemic agent	*Catharanthus roseus*

Source: Lewis H. Ziska, Paul R. Epstein, and William H. Schlesinger, "Rising CO_2, Climate Change, and Public Health: Exploring the Links to Plant Biology," *Environmental Health Perspectives* 117, no. 2 (2009): 155–58.

So, having looked at other aspects of plant biology, let's turn to medicine: How does more CO_2 affect morphine production? Yeah, right, the government is going to allow me to grow morphine. As if.

Sometimes it helps to know something about plants. The commercial poppy used for morphine production is *Papaver somniferum* (not legal in the United States), but there are several dozen poppy species, and, as luck would have it, one other species also produces opiates: *Papaver setigerum*. A weedy relative of *P somniferum*.

And it's legal to grow. Why? Because it produces only a small amount of morphine alkaloids relative to *P. somniferum*. Think of it as hemp to marijuana.

We begin one afternoon when I call up Sini Panicker, a scientist at the U.S. Drug Enforcement Administration (DEA) and tell her what I'm about to do. (This after being transferred to five different individuals and hung up on once.) A moment of silence, then a drawn-out "Coooool."

Scientists are weird. But in a good way.

So we formulate a hypothesis (suppose!). We are going to grow *P. setigerum* at four different CO_2 concentrations—300, 400, 500, and 600 ppm—a pretty good range, from recent to projected concentrations in atmospheric CO_2. And as with the other studies, we focus on the two basics: How much more does the plant grow, and how does its chemistry change? We found that it grew quite a bit. CO_2 concentrations above 300 ppm significantly increased leaf area and aboveground biomass.[38] (CO_2 is plant food!)

Then things got *really* interesting. Reproductively, increasing CO_2 from 300 to 600 ppm increased the number of floral capsules, capsule weight, and latex production by 3.6, 3.0, and 3.7 times, respectively, on a per-plant basis. And the concentration of all opiates—morphine, codeine, papaverine, and noscapine—increased with more CO_2 on a per-plant basis, with the greatest relative increase occurring with the most recent increase in CO_2 (from 300 to 400 ppm).[39]

Whoa.

TWO SIDES OF THE COIN

You might pause at this moment to ask, "If CO_2 can increase morphine production, and morphine is an important pain killer worldwide, then shouldn't this information have fallen into the "good," not the "OMG," category?

Tempting, but no. And to understand why, I will turn to aspirin and its history. Aspirin, or acetylsalicylic acid, was synthesized and branded by the Bayer pharmaceutical company around 1899.[40] However, the pain-relieving properties of salicylic acid had been recognized since civilization began. When the ancient Sumerians weren't fooling around with poppy, they recorded on their clay tablets that willow bark and other salicylic-rich plants like myrtle could be used to reduce fevers, ease pain, and reduce inflammation. The ancient Egyptians noted the medicinal properties of these plants as well, only on papyrus. But they also noted that salicylic acid could cause severe skin irritation.

Around 1897, Felix Hoffmann at Bayer started working to find a less irritating substitute for salicylic acid. He turned to this idea because his father was suffering the side effects of taking sodium salicylate for rheumatism.[41] Hoffman decided to acetylate it, lowering its pH to transfer an acetyl group (CH_3CO-) to the salicylic acid, making it both easier to take and more effective. (And easier to manufacture—no more willow bark!)

Needless to say, Bayer's aspirin was successful, earning the company lots and lots—and lots—of money. The next step was to repeat the chemistry, to acetylate other things. How about morphine? It turns out that when you boil morphine for a few hours, it becomes acetylated, forming a new drug: diacetylmorphine. Gosh, what could Bayer do with this new drug? Well, first it needed a catchier name; "diacetylmorphine" doesn't just jump off the shelves.

So, they named it "heroin" after the German *heroisch*, meaning "heroic" (figure 9.8).[42] And they sold it as a cure for . . . morphine addiction. Honest.

Bayer launched a promotional campaign (prior to doing any research) that touted heroin as nonaddictive and an effective remedy for bronchitis, tuberculosis, and other cough-related illnesses. And they had help. In 1906, the American Medical Association provided its stamp of approval on heroin and suggested it be used in place of morphine. (This continued until the 1920s, when laws in the United States outlawed sales of heroin. Restrictions were then imposed by the League of Nations, and the global manufacture and sale of legal heroin began to decline.) Currently in the United States, between ten and fifteen thousand people die each year from heroin overdose.[43]

So if more CO_2 makes wild poppy grow more and makes it produce more morphine and other opiates, could more CO_2 mean more heroin from commercial poppy?

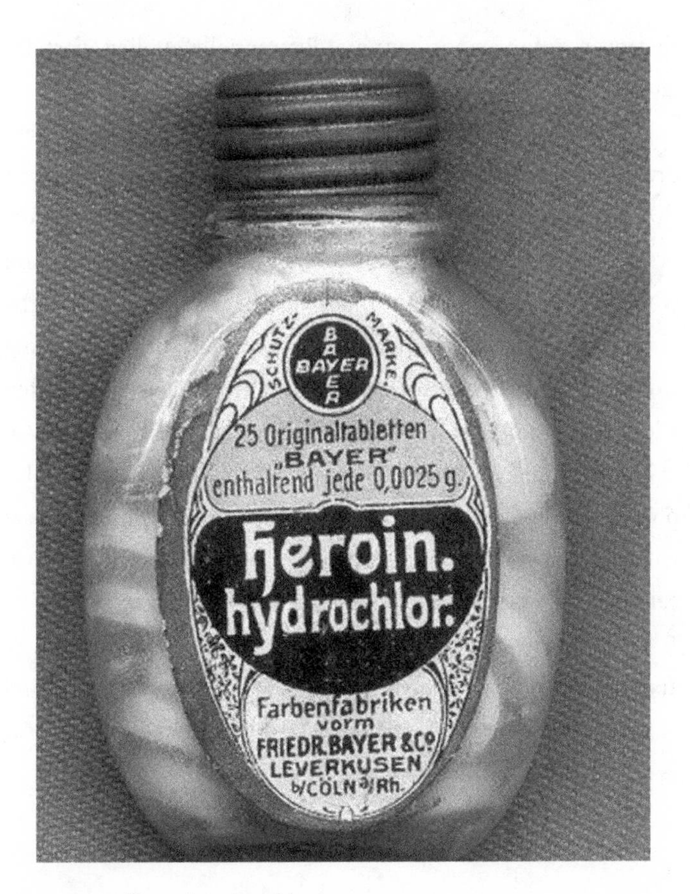

FIGURE 9.8 Bayer heroin tablets.
Royal Pharmaceutical Society Museum, London, public domain.

The "war on drugs" has been ongoing for fifty years—perhaps not so much a war against drugs per se but against the classes and culture associated with certain drugs. The first opium laws against Chinese immigrants were established in the 1870s, and the first anti-marijuana laws were brought against Mexican migrants in the early twentieth century. There is little question that drug enforcement laws are differentially applied, particularly between Black and white communities.

But the "war" should include some aspect about growing these drugs and limiting supply. And as you might imagine, the work that I did with Sini at the DEA showing how CO_2 could affect opiate production in wild poppy might be, well, interesting. And what about other narcotics? For decades, Colombia and the United States have engaged in a war on drugs, over everything from marijuana to heroin. But one drug in particular has gotten a lot of attention: *Erythroxylum coca*, or cocaine. Can rising CO_2 also affect cocaine production, so that even if spraying with glyphosate (Roundup) destroys part of the crop, the remaining part will produce more of the narcotic? We've already seen how CO_2 can alter the effectiveness of herbicide on other "weeds."

If the DEA were really interested in how CO_2 could affect where and how narcotics are grown, they might want to support this kind of work. It would be fundamental to any war on drugs. But as near as I can tell, the DEA has shown zero interest. Publicly anyway.

ONLY GOD CAN MAKE A TREE

The "CO_2 is plant food" meme is effective in part because it can be so misleading. When you think "plants," heroin isn't the first thing to pop into your mind. Perhaps you think of something statelier, an oak perhaps or a giant sequoia. It isn't a coincidence that when the meme is mentioned, especially on conservative websites, a loblolly pine tree is shown growing more in response to increased CO_2.[44] (If you're curious, the CO_2 Coalition is funded by the

Mercer family and run by William Happer, the well-known physicist who likened CO_2 promotion to Holocaust denial.)[45]

But more CO_2 means more trees, and that's a good thing, yes? Hmm, could be. Depends, of course, on the tree. William Laurance is a former Smithsonian scientist who for many years has been looking at climate and the impact of rising CO_2 on tropical forests. He has documented that, yes, more CO_2 can increase tree growth. But also that faster-growing trees were benefitting at the expense of slower-growing subcanopy plants.[46] As pointed out earlier, adding a resource does not mean that all species will respond in the same way. In this case, if only a select number of tree species respond to CO_2, biodiversity will be affected, impacting all animal life, from insects to jaguars. Good for some trees, not so good for the forest.

PLANTS AND AIR POLLUTION

This idea may seem a bit weird, in part because we think of plants as air "cleaners" filtering the air in a room, but they can and do release chemicals into the air that have implications for human health. Some of these are obvious; for example, invasive plants such as cheatgrass can contribute to fires; fires that produce ash and smoke and have indirect consequences for air quality.

But there's more.

If you are from the South, you are likely familiar with kudzu, an invasive vine that is one of the fastest-growing plants on Earth (and featured on the book's cover). A number of epithets describe kudzu, including "a vegetative form of cancer" and "the plant that ate the South." I think of it as that monster plant in old black-and-white science fiction movies from the 1950s whose tendrils snake into the open windows of houses to abduct unsuspecting inhabitants. (I'm convinced the carnivorous triffid plants in the 1962 movie *The Day of the Triffids* were based on kudzu.)

How fast does kudzu grow? At present it covers about eleven thousand square miles in the United States, and, while estimates vary, it may colonize up to two hundred more square miles each year.[47] As with cheatgrass, once it gets going, it wipes out other plant species, dominating the landscape and diminishing species diversity (figure 9.9).

It's well recognized that when insects land on and chew leaves, some plants release volatile organic compounds (VOCs)

FIGURE 9.9 Kudzu, an invasive vine found throughout the southeastern United States.

U.S. Department of Agriculture, Agricultural Research Service, public domain.

into the air, compounds that are "detected" by other plants who then ramp up their production of chemical "weapons" in response. However, VOCs are also a source of air pollution.

And kudzu produces two key VOCs.[48] The first is isoprene, released from the leaves. The second comes from the roots, which convert atmospheric nitrogen into ammonia, converted in turn by soil bacteria into nitric oxide. If sunlight is present, isoprene and nitric oxide mix to make ozone. (Nitric oxide is also a powerful greenhouse gas.) Kudzu isn't alone in making VOCs, but it seems that kudzu can produce more isoprene and nitric oxide than other plants, with consequences for ozone, a ground-level pollutant, which because of its oxidizing potential can cause damage to respiratory tissue.

Oh, did I mention that several studies indicate that more CO_2 induces a strong growth response in kudzu?[49] The implications for its ability to dominate the landscape are obvious, but it seems possible—even probable—that more CO_2 may also contribute to increases in ground-level ozone.

FIRE IN THE BOLE

Wildfires. Unless you've been living on planet Xenon, the link between climate change and wildfires is something you've heard or read about. But what does rising CO_2—by itself—mean for wildfires?

Not surprisingly, large fires are driven mostly by plant material. And if more CO_2 means more plant material, then that is more fuel for the fire. (CO_2 is plant food!)

What if an increase in CO_2 alters plant (tree) chemistry in a way that leads to changes in wildfires? What if more CO_2 make trees more flammable? Or makes trees burn hotter? If you recall, we looked at cheatgrass (an invasive weed throughout the western United States) and saw that recent and near-term future CO_2 levels were already having these effects. We cannot presume the same for trees of course, but if cheatgrass is any guide, we need to look into it. So researchers are looking into if or how an increase in CO_2 affects tree chemistry and flammability. If so, what does that mean for the kind of smoke that might be produced?

That last bit isn't trivial, as smoke affects air quality and all living things that need air to breathe. Wildfire smoke has all kinds of less-than-wonderful qualities, from carbon monoxide to nitrogen oxides to fine particles. These qualities may explain why breathing in smoke can lead to respiratory- and cardiovascular-related emergency department visits, asthma, bronchitis, chest pain, chronic obstructive pulmonary disease, respiratory infection, and long-term lung illness. And, not surprisingly, hundreds of thousands of deaths annually. It also exacerbates the effects of COVID-19.[50] So if CO_2 by itself has any effect on this process, it could be a big deal.

These examples are by no means the whole story, and there is much, much more to be learned. Yes, it is easy to say that CO_2 is plant food and to imagine that a plant growing more is "good." But when you begin to understand that not all plants respond in the same way to CO_2 and that increasing CO_2 will cause a global shift in plant communities, from crops to weeds to trees to vines, then you can begin to grasp how *all life* will be affected by increases in atmospheric CO_2. And that merits an OMG.

PART III

CO_2 IS PLANT FOOD. NOW WHAT?

CO_2 INTERACTS with plants in ways that will impact many aspects of human society, from nutritional quality to allergies to, well, just about anything. But it's doubtful you will have heard about these OMG CO_2 effects because—say it with me: "CO_2 is plant food."

As with any assessment, I am sure to have missed something fundamental about plant biology and CO_2 that falls into the good, bad, or OMG category. Yet the bottom line is simple: CO_2 will significantly alter all ecosystems on Earth because plants make up those systems. Because when a resource like CO_2 increases, you can't just look at one plant and say, "Wow, this is great!" because not all plants respond equally—and that disparity will alter the composition and vitality of the world's ecosystems. For the managed systems, like agriculture, weeds appear to be winners and crops losers. For the unmanaged, natural systems like rainforests, vines and select tree species may be the winners, with biodiversity, including the animal species of rainforests, the ultimate losers. CO_2-induced changes in plant chemistry also can't be ignored. Such changes will affect every aspect of our daily lives, from the nutritional quality of our food to plant-based medicines. For ecosystems, changes in the ability of plants to ward off insect predators and the ability of bees to obtain proper nutrition will disrupt nature in ways that have not yet been considered. Altering plants—how they grow and how they function—might seem kinda important. At least one would think so.

But. The politics, the endless repetition of "CO_2 is plant food" by conservatives, and the inroads that message has made on the American (even global) psyche means that we aren't studying the CO_2 effect. Or at least not very much. For example, if you look at the 2019 National Institute for Food and Agriculture (NIFA), the institute that supplies about a billion dollars in grants to agriculture universities in the United States, there is no mention of CO_2. None. For that matter, the phrase "climate change" isn't included either, except for a few euphemisms like "climate uncertainty."[1] Applying for funding also often involves running into a catch-22. For example, let's say we want to get research funding to study the effect of CO_2 on pollen and allergies. We apply to the National Science Foundation (NSF), a primary funder of science. Their response: "This is a health issue. You should apply to the NIH (National Institutes of Health)." So we apply to the NIH. Their response: "This is an environmental issue. You should apply to the NSF." Who knew federal agencies were like parents responding to a sleepover request?

So far, I've provided an explanation of the low-hanging fruit, the *cheap*, low-hanging fruit of the "CO_2 is plant food" issue, and hopefully you've gotten a sense of just how important CO_2 by itself can be. But when you're dealing with a gas that affects all life on the planet, there are consequences—primary, secondary, and tertiary. As with any magic trick, simplicity is an illusion. So let's dig a bit deeper. What do we know? Why don't we know more? What needs to be done?

Chapter Ten

MORE QUESTIONS THAN ANSWERS

I HAVE presented several examples of the good, bad, and OMG with regard to increasing atmospheric CO_2 and plant biology. Other possibilities remain speculative for now because we don't yet have the data we need. But some are worth mentioning because they are of *potential* consequence in the "CO_2 is plant food" arena and as such warrant further investigation. And funding.

As previously mentioned (ad nauseam by this point), "CO_2 is plant food," and any rapid increase will differentially affect plant species, some will win, while others lose. Still, if they respond differently, then doesn't CO_2 become a selection factor, an *evolutionary* driver? How might that work?

Let's go back to weeds and herbicides. The number of weeds resistant to herbicides is increasing rapidly (see figure 8.3), posing a unique threat to farmers and crop yields. But a given weed species often has two biotypes: sensitive and resistant. The sensitive biotype remains vulnerable to the herbicide, the resistant biotype not so much. So as CO_2 rises, will it select among these biotypes? Good question.

Sadly, we don't yet have an answer. Just one study so far. But this study indicates that both higher temperatures and more CO_2 favor the resistant biotype of an important weed in rice: junglerice (*Echinochloa colona*). So more CO_2 may select for that biotype.

Suppose (!) that this study is representative. Imagine what that means for crops and weeds. Imagine how much more difficult it will be to control weeds in the future if more CO_2 favors resistant lines. Imagine what that means for crop production. As weeds pose a greater and greater threat to yields, and herbicides become less effective, what will that mean for our ability to feed the eight billion people of the world?

Let's look at another common CO_2 effect: reducing the nutritional quality of plants. Most of what we know is based, understandably, on what humans eat. The study on goldenrod pollen discussed in chapter 9 is the exception. But again, let's suppose that one study is the norm.

Many of us learned about food chains in high school biology. One thing eats another thing, which in turn is eaten by something else. At the bottom of the food chain are plants—the only living things that convert sunlight into chemical energy. And at the next step are the largest consumers of plants: insects. Insects are the linchpin of animal food chains. They are essential food for birds, bats, reptiles, amphibians, and fish, and they are vital to ecosystem function, from pollination to pest control to nutrient recycling.

Insects outweigh all the fish in the oceans and all the human beings on land. As of the last count, there are about five to ten million species, with more being found almost daily.[1] They interact with plants in a fascinating evolutionary dynamic that words cannot describe, but examples range from pitcher plants imprisoning, ingesting, and digesting insects to orchids that simulate the sexual movements of butterflies with their long, weaving stalks to attract pollinators (who then copulate with the flower).

In recent years, you may have come across the term "insect apocalypse," the sudden, dramatic decline in insect populations occurring across Europe and North America.[2] One can debate just how vast this decline is and whether the world will soon run out of insects (except for cockroaches, let's be fair), but the honest answer is that we aren't sure. Not just because of the insects but also the tumbledown consequences of their disappearance. As reported in the *New York Times*, birds that rely on insects are also declining.[3] There has been an 80 percent drop in turtledoves and a 50 percent drop in nightingales, and half of all farmland birds in Europe have vanished in the last three decades.

Why? Pesticide use seems obvious, and habitat destruction is another likely cause.

But what about the effect of increasing CO_2 on the nutritional quality of food? Remember the work on CO_2 and goldenrod, which found that recent increases in CO_2 have reduced the protein content of goldenrod pollen by 30 percent. What if this is a general effect, if some critical nutrition threshold has been reached as CO_2 has risen, and insects are literally starving to death owing to a lack of protein?

Seems far-fetched. Yet Ellen Welti, while a postdoc in the Konza Prairie Long-Term Ecological Research program in Kansas, looked at grasses consumed by grasshoppers and discovered something interesting.[4] The nitrogen and phosphorous concentration of prairie grasses has declined in recent decades, suggesting that a CO_2-enriched world could directly contribute to declines in insect herbivores by affecting their nutritional intake.

So any global effect of more CO_2 on the nutritional quality of food may also be affecting a key function of the food chain—in addition to habitat loss and pesticide use. And if more CO_2 results in a global decline in nutrition for insects, the amount of life the earth can support will decline. All of it, the birds and the bees, you and me. So many questions—important ones—are swept aside when people say, "CO_2 is plant food."

Chapter Eleven

THE TEN-TON *T. REX* IN THE HALL CLOSET

QUICK PASSING of a colleague in the hall.

"Hey, how's the rice model on CO_2 and temperature going?" I ask.

"It's done," is the response. I stop and turn to him as he keeps moving down the hall.

"Wait. What?" I yell. "We just got the grant a year ago." I move to him quickly.

He stops and turns, a hint of a smirk, "We already have temperature models. We just adapted them for rice; it wasn't hard. The final report will come out in a couple of weeks."

Still incredulous, I put my hands on my hips and ask, "What does CO_2 do to the temperature response?"

The scorn is more evident as he walks away waving a hand in dismissal. "They're separate issues." (Unfortunately, this is still the opinion of many scientists when it comes to plant biology.)

That view in part reflects methodology. As mentioned, FACE (Free-Air CO_2 Enrichment) is often seen as the best means to expose plants in the field to elevated CO_2, mostly because it doesn't involve an enclosure. However, as you might imagine, elevating temperatures over the large area of a field without an enclosure is really HARD. Hence, much of the available FACE data focus only on CO_2, not temperature. It wasn't that CO_2 and temperature data didn't exist—but recall that FACE was the gold standard—although comparisons among FACE and other means of applying CO_2 are more sophisticated than might appear at first.[1] Still, these data also fit into existing political memes—CO_2 is plant food (yay!) but CO_2 means climate change (higher temperatures) (boo!). All good, or all bad.

I wrote this book to look at CO_2 and plant biology, but it would be silly not to mention the ten-ton *T. Rex* in the hall closet: the role that CO_2 and other greenhouse gases play in increasing surface temperatures. What is the impact of the greenhouse effect (or climate change or global warming or whatever euphemism you think appropriate) on "CO_2 is plant food"?

So much has been written on the issue of climate change that to repeat it here would be self-defeating. If you would like to know more, I recommend the work of David Wallace-Wells[2] and Naomi Klein.[3] The best I can do is to take Inigo Montoya's approach in *The Princess Bride* and say that there is too much, let me sum up.

Here is my attempt to bypass semesters of chemistry and atmospheric physics to explain climate change. Let me borrow your guitar for a moment. I will tune two strings side by side to the same frequency; let's say A. Now I will pluck one string; what will the string next to it do?

If you said vibrate or resonate, slap a star on your forehead. If the strings are tuned to the same frequency, that second string will absorb some of the energy of the first string, energy that would otherwise escape. This is resonance.

And that's what water and carbon dioxide do in the air. They don't resonate in the key of A, of course, but in the key of infrared, or heat. When these molecules encounter heat, they resonate—vibrate—and absorb some energy that would otherwise be lost.

This is a good thing (honest). Otherwise, the earth would be much cooler. (The average temperature would be about −18°C, not the current average of about 14°C, soon to be 15°C). Water vapor and carbon dioxide create a natural greenhouse effect that makes life (much of it anyway) possible.[4]

But you can see the dilemma, the Goldilocks paradox: too few greenhouse gases and you get cold (Mars); too many and you get hot (Venus).

There is another irony—water in the air (relative humidity) and carbon dioxide are not increasing together—only CO_2 is going up. So the earth is warming differentially. That is, where it's warm and wet (e.g., the tropics), where water vapor (humidity) already represents a "greenhouse" effect, more CO_2 will warm things up—a bit; but, where the air is dry (e.g., the poles, deserts), where water vapor is low, adding CO_2 will have a greater relative effect on temperature. And, in general, this is what we see—greater temperature increases at the poles relative to the equator, increased desertification—hotter deserts, warmer winters relative to summers, etc. This differential warming will, as you can imagine, cause some extreme weather as well.

I know, I skipped a bunch of important stuff in that explanation, but it's the gist of climate change. That is, the impact of rising CO_2 on warming will depend (roughly) on how humid a region is to begin with. Let's take this quick lesson and ask, for good or ill, the question my colleague ignored: What impact will rising CO_2 *and* climate change have on plant biology? What are the consequences?

Chapter Twelve

WAIT, WHAT?

SO NOW we have two major impacts: the direct effects of CO_2 on plant biology and the indirect effects of CO_2 on temperature and climate. But are they really independent, as my hallway friend asserts?

Would that mean that one ("CO_2 is good!") could cancel out the other ("hotter temperatures are bad!")? Isn't that what happens when you split the difference? There is a null effect—they cancel each other out.

But what if one exacerbated the other? What if rising CO_2 didn't negate the effect of rising temperature on plant biology . . . but made it worse. How would that happen? CO_2 stimulates plant growth, as does temperature, so shouldn't they act in sync, making CO_2 an even more effective growth stimulant?

One of the most cited papers in the "CO_2 is plant food" universe (more than 1,100 citations since 1991!) is an overview by Stephen Long of the University of Illinois entitled

"Modification of the Response of Photosynthetic Productivity to Rising Temperature by Atmospheric CO_2 Concentrations: Has Its Importance Been Underestimated?"[1] Long states,

> An increase in C_a from 350 to 650 µmol mol–1 can increase A_{sat} by 20 percent at 10°C and by 105 percent at 35°C, and can raise the temperature optimum of A_{sat} by 5°C. This pattern of change agrees closely with experimental data. At the canopy level, simulations *also suggest a strong interaction of increased temperature and CO_2 concentration* [my emphasis].

"A_{sat}" refers to photosynthesis, the basis for obtaining the carbon necessary for plants to grow. Long is advocating that you can expect a strong synergy between temperature and CO_2 when it comes to stimulating photosynthesis: Warmer temperatures may result in a five-fold increase in photosynthesis at higher concentrations of CO_2 (the "C_a" in the quote). The clear implication, Long explains, is that we (the scientific community) are underestimating how much temperature and CO_2 will act together to stimulate plant growth. In simple terms, if CO_2 is plant food, the amount of food available should go way up as it gets warmer.

A fascinating finding. And one that was taken as the final word on the topic by the scientific community for many years. But it turns out to be more complicated (shock!). As it happens, plants consist of more than just leaves (which was the basis for Long's assessment of photosynthesis). They also consist of seeds and fruit—and pollen, which once again provides an interesting contrast.

TABLE 12.1 Optimal temperatures for different crops

Crop	Optimal Temperature, Vegetative	Optimal Temperature, Flowering	Failure Temperature, Flowering
Cotton	34°C	25–26°C	35°C
Maize	28–35°C	18–22°C	35°C
Peanut	31–35°C	20–26°C	39°C
Rice	28–35°C	23–26°C	36°C
Sorghum	26–34°C	25°C	35°C
Soybean	25–37°C	22–24°C	39°C
Wheat	20–30°C	15°C	34°C

Data adapted from Jerry L. Hatfield, Kenneth J. Boote, Bruce A. Kimball, L. H. Ziska, Roberto C. Izaurralde, D. R. Ort, Allison M. Thomson, and D. Wolfe, "Climate Impacts on Agriculture: Implications for Crop Production," *Agronomy Journal* 103, no. 2 (2011): 351–70.

Pollen is sensitive to temperature. Optimal temperatures for pollen are 5 to 10°C less than those for leaves (table 12.1). So as temperatures go up, pollen viability goes down. Plants can become sterile. Could this affect photosynthesis—and, more to the point, the ability of more CO_2 to stimulate seed yield?[2]

Photosynthesis, which is how plants use light energy to capture carbon (as CO_2), is sink dependent; that is, there must be a need for carbon, a sink or reservoir where that carbon ends up. If sinks are limited or nonexistent, then photosynthesis—which has a metabolic cost—backs up and slows down. This is usually referred to as feedback inhibition.

As you might guess, one of the biggest carbon sinks are developing seeds, or fruit—they represent a big investment of carbon—as proteins, sugars, or carbs. And it's a necessary sink: no seeds, no reproduction. But if something affects that sink—say, warmer temperatures—then pollen can become sterile, and fruit *doesn't* set—and carbon sinks diminish. And if sinks diminish, so does photosynthesis—it backs up—*as does any response to more CO_2.*[3]

The idea that temperature and CO_2 will consistently act in sync in a synergistic way to promote photosynthesis and growth is not, contrary to Long's finding, what happens for the whole plant. Rather, there is a balance, one that temperatures influence.

Indeed, since Long's original proposition that temperature was being underestimated in terms of CO_2 stimulation, few studies have confirmed the idea at the level of the whole plant or field. On the other hand, many studies have suggested that temperature can and does negate any positive benefit from CO_2.[4]

How? One way is what we just discussed: Temperature restricts carbon sinks, photosynthesis backs up, and the plant can no longer take advantage of any additional CO_2. But there is another physiological consequence of CO_2 that deserves a bit more attention.

We know that plants "breathe." Specifically, they lose water from their leaves in exchange for taking up CO_2. It is a very unequal exchange. Most plants lose about one thousand water molecules to gain a single molecule of CO_2. So as CO_2 increases, stomata (the pores in leaves that exchange water for CO_2) close in response. If there is more CO_2, stomata don't need to open as wide, and the exchange rate of water loss for

CO_2 gain becomes more favorable. But there are consequences. Less water being lost means less evaporative cooling. Plants heat up. The area around the plants, the canopy, heats up. And the pollen heats up. And if both temperature and CO_2 go up, yields may decline even more than they would with warmer temperatures alone.

Figure 12.1 illustrates data from a field experiment on rice in the Philippines.[5] The "percentage of filled spikelets in rice" label on the y-axis refers to the number of seeds that developed following pollen fertilization. Values between 90 and 100 percent mean that the pollen was fertile and 90 to 100 percent of the

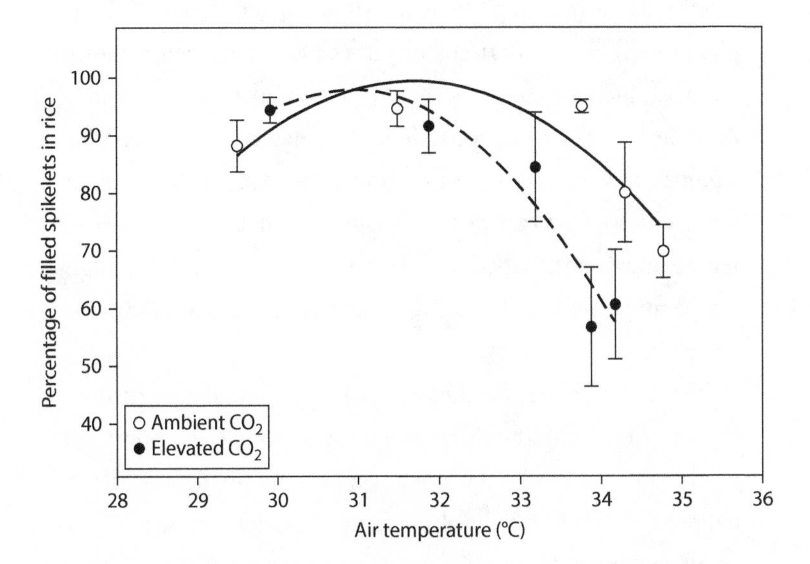

FIGURE 12.1 Change in spikelet sterility in rice as temperature increases at current and future CO_2 concentrations. At elevated CO_2 concentrations, pollen is more sensitive to increasing air temperatures.

Figure by the author.

seed was set. But as the temperature increased, viable pollen—and thus seed set—declined. Not surprising.

But what is interesting, and what the data in figure 12.1 show, is that seed set declined faster at the elevated CO_2 concentration (650 ppm) as air temperatures rose. Why? Because CO_2 reduced the size of the leaves' pores, meaning the amount of cooling was reduced. When this happens, plants experience a higher temperature even if the air temperature has not changed. The result? A greater risk of pollen sterility and, in the case of rice, the world's most important food crop, greater vulnerability to temperature extremes—extremes that will increase significantly as average temperatures go up.[6]

CO_2 can stimulate photosynthesis and plant growth. (CO_2 is plant food!) But study after study has shown that if temperature and CO_2 increase together, there is no stimulation—there may even be a decline—in seed yield. This has been observed for legumes, wheat, sorghum, rice, and corn.[7] CO_2 may be plant food, but so far the evidence suggests that CO_2 exacerbates temperature sensitivity. So at the end of a season with high temperatures and high CO_2, you may have a big leafy plant. But no seeds.

These results are antithetical to Long's initial observation. But as with many things in science, perspective is important. Looking at the leaf level is not always a good predictor of what happens at the flower level; looking at flowers doesn't tell us about root biology; root biology doesn't necessarily tell us about what the entire plant is doing; and what the plant is doing doesn't predict how the ecosystem might respond, and so on.

It is possible that at certain temperatures, CO_2 has no effect on crop yield. But again, we need more information before we can make such a broad conclusion. What is clear is this: Temperature and CO_2 are *not* separate issues. As Long's 1991 paper illustrates, it is not enough to study only one level of organization (in this case, leaves) and assert that you know everything. Studying photosynthesis is useful, but scaling it to a global, one-size-fit-all response without doing the necessary research is problematic. We need to know more, much more.

Why don't we?

PART IV

POLITICS AND PLEAS

Chapter Thirteen

CRACKS IN THE SYSTEM

THERE ARE many appeals to address climate change. From student protests and class absences to boycotts, street demonstrations, videos, speeches, social media, and tweets from movie stars and politicians. But to my knowledge, none of these pleas focuses on the role of rising CO_2 on plant biology globally. There appears to be little desire on the part of the public to want to know more. Why?

This is where the introduction and proliferation of the "CO_2 is plant food" meme by the fossil fuel industry has been nothing short of brilliant. Their use of videos such as *The Greening of Planet Earth* (guilty as charged for my part in this if you recall the preface). Their endless PR campaign to find a simple phrase that, through endless repetition, succeeds in becoming a global meme.

And what makes it particularly egregious is that the phrase has elements of truth. It's hard to deny that CO_2 is plant food.

It is. It's hard to deny that CO_2 stimulates plant growth. It does. It's hard to say that stimulating plant growth is bad. Trees are good! Let's hug one, you hypocritical liberals!

Attempts to respond to "CO_2 is plant food" with short, catchy phrases is difficult. "Well, it depends . . ." doesn't cut it. "CO_2 is plant food" makes a great bumper sticker. "Any beneficial plant response will be miniscule relative to the complete devastation climate change will impose on Earth systems" doesn't. Unless you have two cars traveling in tandem.

Memes are powerful. To illustrate why, let me recall a visit to Capitol Hill some years back. I was there to talk about climate change and weed biology, to provide some background on crop production and climate change to a member of the House Agriculture Committee (a "small-d" Democrat).

Visiting Capitol Hill involves talking to staffers, the heart and soul of how Congress functions. While presenting my research, I was told by a staffer that the congressperson could see no reason to increase funding for agricultural research on CO_2. They had access to Google, the "internet library," and what they had googled so far indicated that CO_2 was good for plants.

Nothing against Google, but the last time I checked there was no "truth-o-meter" for a given website. But if we go a step further with that logic, why would anyone want to fund science? Who needs new research? Let's just make stuff up; let's "post-truth" it. What are a few "alternate facts" between friends?

Knowledge is important. Critical thinking is essential. Asking questions is critical. Why? Because they are necessary to sift fact from fiction. Otherwise, you can only "feel" if someone is telling you the truth. This is why we propose hypotheses in science: Suppose we did this? What happens if we try that? We want proof—evidence. And we are justifiably skeptical if anyone in power says " 'Cause I say so."

Significant and powerful interests want very much to insist in no uncertain terms, with evangelical flair amid blaring trumpets, that "CO_2 is plant food" and that more CO_2 is a benefit to the planet, a herald of a new Eden. And they are projecting that catchphrase into the political morass of the climate change conversation.

Who are these prophets? To get a closer view, let's take a snapshot from May 2019 of a congressional hearing on a UN report on biodiversity.[1] This report indicated that up to a million species were at risk of extinction in response to climate change. Republicans invited two "experts" to comment. One was Marc Morano, a former producer for the noted environmentalist Rush Limbaugh and now the communications director for Senator James Inhofe of Oklahoma (who gets a lot of money from the oil and gas industry from that state—what a surprise!). Marc has no scientific degrees but does have a BS (how appropriate!) in political science from George Mason University. He is considered the "Matt Drudge of climate denial." During testimony, he stated, "Humans putting carbon dioxide into the environment are the salvation of life on Earth."

The second "expert" was Patrick Moore, who does have a science degree, a PhD in ecology from the University of British Columbia, and has a PR firm that has represented the mining industry, the logging industry, the nuclear industry, PVC manufacturers, and biotechnology. (Hey, he's diverse). His comments on CO_2? "So I say celebrate CO_2. It is the most life-giving substance along with water on this planet, and it's doing the world a lot of good."

See? More CO_2 in the atmosphere? It's all *good.*[2] One doesn't have to go far to find the "CO_2 is plant food" talking point repeated. But perhaps the most over-the-top, summa cum verbiage has come from William Happer, a senior director at the National Security Council and an emeritus professor of physics at Princeton University: "The demonization of carbon dioxide is just like the demonization of the poor Jews under Hitler. Carbon dioxide is actually a benefit to the world, and so were the Jews."[3] Uh-huh. Who knew that being against "CO_2 is plant food" is the same as committing genocide? And the response from across the political aisle?

Meh. CO_2 isn't going to do anything. Why? Because it depends on other aspects of climate change or temperature or nutrients or, well, . . . something, something.[4] The bigger issue is climate change—melting ice caps and drowning polar bears. Yes, CO_2 is plant food, but don't look here; look over there.

Politically, it is made to seem like a black-and-white issue. Scientifically, of course, it is complex. Complex in a way that deserves more than political entrenchment of the "You're wrong, I'm right!" variety.

Chapter Fourteen

SCIENCE SAYS

ONE WOULD logically suppose that given that life on Earth is dependent on plant biology and that all plant species are, or will be, impacted by the ongoing rise in global CO_2 concentrations, a concerted effort would be made by *all* government research agencies to understand and address this issue. This could include the Centers for Disease Control and Prevention studying the implications of increased CO_2 on food allergies, or the Food and Drug Administration evaluating the impact of increased CO_2 on plant-based medicines, or the National Institutes of Health working to understand more about the impact of increased CO_2 on nutrition and plant-based allergies, or the Drug Enforcement Administration researching how increased CO_2 affects the production of narcotic plants. One would suppose incorrectly. No effort is being made.

One could even hope that conservative politicians who have been repeating "CO_2 is plant food" would be interested. I mean, if CO_2 is the greatest thing since canned beer, then besides promoting it in climate change discussions, they might ask for additional research on its impact. Maybe conservatives could learn from marijuana growers and stimulate the economy by investing in the most CO_2-responsive varieties of ganja (thank you, John Boehner!).[1] Don't they want to help the American farmer grow the most CO_2-responsive lines of wheat? Or apples or whatever? Shoot, they could do it just to satisfy their own egos, to be able to point to the research and say, "See? CO_2 is plant food!" and then turn to liberals, middle fingers extended.

Of course, none of this has happened. Why? Because "CO_2 is plant food" is not a scientific argument but a political assertion wrapped in "green is good" assumptions—and, unfortunately, one that the public has swallowed hook, line, and pine tree.

Yet scientific inquiry conducted as part of a political agenda doesn't sound like science to me. It doesn't matter whether it's right, left, or center politically; any results driven to *confirm* what you believe cannot be trusted. And if there is one truism to which all politicians should pledge allegiance, it's that sound science ought to be the basis of policy-making. Science first, then politics. Doing it the other way around destroys democracy. And obliterates reality.

The good, the bad, and the OMG that I've described in this book are based on observed and potential responses of plant biology to increasing CO_2. And these responses are or will be massive—and separate from those of global warming. This fact is not part of the current mindset, not part of the public discussion, and, with few exceptions, not part of ongoing scientific research. I have nothing against studying sea level rise, hurricanes, or drowning polar bears. There are clear connections between climate change and these phenomena. These issues are important, global, and will affect millions of lives.

But plant biology is also responding to rising CO_2 levels. I hope that the previous chapters have been instructive. Plants are essential, fundamental, necessary to every living thing on Earth. Indispensable to human existence and to civilization. If you change a fundamental resource that plants need to grow, *all life will be changed.*

And you cannot—*you must not*—assume that this change will be uniformly beneficial. "CO_2 is plant food" is not a shiny trope that excuses or offsets climate change consequences. That is the goal of climate change deniers. Based on the data we have, a more Eden-like environment seems an unlikely outcome. If, as we saw with farming systems, more CO_2 leads not to a uniform increase in the growth of all plant species but to an increase only in those with greater genetic variation and greater environmental adaptability (i.e., weeds), then there will be winners and losers within all ecosystems.

That last bit is important. Yes, more CO_2 can make plants grow more, but it will *not* make all plants grow equally. And that differential response will affect species diversity, plant competition, plant chemistry—and ultimately evolution. Not only of plants but of all living things, including us.

Conversely, liberals cannot dismiss or ignore the "CO_2 is plant food" meme. It is a fundamental biological truth that plants need CO_2 to grow. Plants evolved at a time of higher CO_2 than today, and as atmospheric CO_2 creeps up, they will respond. And it is imperative to understand that response.

When looking at the impacts of rising CO_2, plant biology—not sea level rise—should be the uppermost of everyone's concerns.

But at present, sadly, the issue of CO_2 and plant biology is falling between the ideological cracks. Little research has been done, and we are awash in ignorance. As with climate change, "CO_2 is plant food" should be a scientific issue, not a political one. But it is not seen by the public as such. To understand why, I have provided the following rough transcript of an actual phone interview with an Iowan farmer. (I thank Eugene Takle, a professor at the Department of Agronomy at Iowa State University, for taking the time to describe the interview to me during an informative phone conversation.)

INTERVIEWER: Are you seeing wetter springs and uncertain, extreme weather?

FARMER: Oh, absolutely. We are seeing shorter and shorter weather windows for planting and harvesting, more storms, more flooding.

INTERVIEWER: Are you seeing new weeds, insects, and disease?

FARMER: Yep, someone said they saw kudzu on their property! And miscanthus. Serious invasive weeds. Not good.

INTERVIEWER: Are you investing in new infrastructure?

FARMER: Big time! New planters so we can plant faster while the weather is good, new tile drains to help combat flooding.

INTERVIEWER: Is this climate change?

FARMER: Oh no, that's Al Gore.

Let me make a plea. I know that science isn't perfect, but Al Gore didn't invent climate change as some sort of get-rich-quick scheme. (If he did, I'm still waiting for my check.) If you have evidence that CO_2 is not a greenhouse gas absorbing heat in the infrared region of the electromagnetic spectrum, let's see it. If you have evidence that CO_2 promotes the growth of crop plants exclusively and not their weedy neighbors, let's see it. Otherwise, keep reading.

It's easy enough to say that scientists are biased because science is what they always turn to. But science is still one of our best means of expressing hope and optimism. Our need, our instinct to explore the world around us is a fundamental aspect of humanity, as defining as art, literature, or music.

And it gets the job done—eventually—because, as with all human endeavors, we make mistakes. But the core of science is skepticism. If you are a scientist reading this, you know—*you know*—that when you present your research at a conference, you will be questioned. That is how truth is discovered. It doesn't happen with the publication of a single research paper but over time, over questions, over challenges.

If you think something is wrong, challenge it. But don't challenge it because it doesn't conform with your religious views, your political leanings, or your feelings. Challenge it with your own hypothesis, your own experiments. Write up what you did, and let others check your work. If they can repeat it, maybe it's worth pursuing.

Science does not have a perfect track record, but it can self-correct. (Remember cold fusion? No? Good. Thank science.) When faced with challenges in the past, science has dealt with them: Electricity from sunlight? Check. A man on the moon? Check. A cure for polio? Check. Science is a practical means of researching, analyzing, and addressing problems. We use science every day, from aspirin to clocks, from doorknobs to cars, from microwave ovens to airplanes.

When it comes to science, agriculture is no exception. Indeed, agricultural research was the core science at the heart of many land-grant colleges (it's the "A" in the names of many institutions like Texas A&M and North Carolina A&T). And it's the reason the United States is still considered a "breadbasket of the world."[2]

At the federal level, two major agencies fund agricultural research: the Agricultural Research Service (ARS) and the National Institute of Food and Agriculture (NIFA), both under the auspices of the U.S. Department of Agriculture (USDA). (This is, frankly, an odd arrangement, as the bulk of federal dollars spent on scientific research comes from the National Science Foundation and the National Institutes of Health. But because of our agrarian past, agricultural research has always been kept separate.) The ARS is the "in-house" research arm of the USDA; NIFA, in contrast, is the primary funding arm for university research on agriculture.

If hard-core conservatives are correct about climate change being a ruse to steal tax dollars, then a Niagara Falls of money must be flowing to federal agencies that deal with climate change. It must be off the charts. Let's see about that. Figure 14.1 provides data for all federally funded agricultural research (not just on climate change) from 2001 to 2018 (in 2001 dollars).[3]

Wow. The amount of money spent on agricultural research is just . . . sad. Not only is nothing being spent on studying the impacts of rising CO_2, nothing is being spent on *any* of the

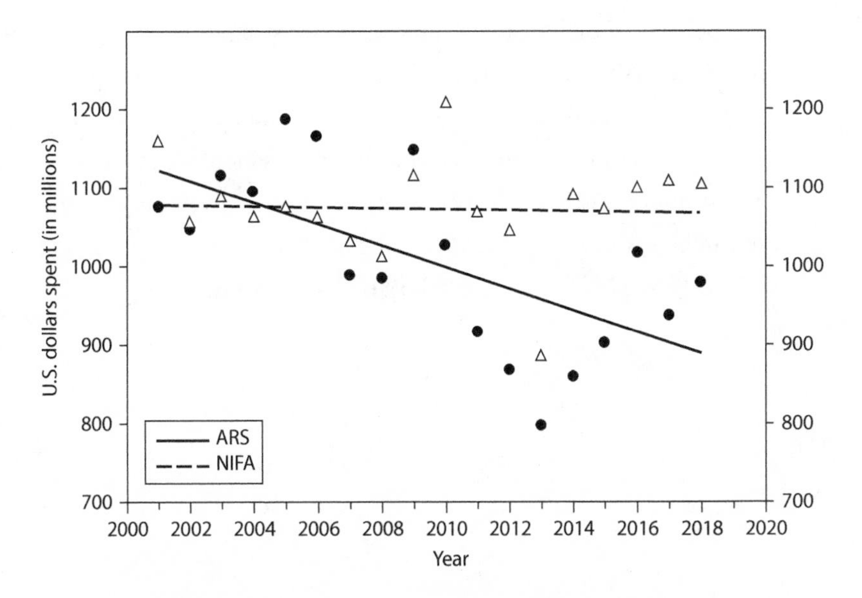

FIGURE 14.1 Budget for the two primary U.S. agricultural research agencies, the Agricultural Research Service and the National Institute of Food and Agriculture (in 2001 dollars).
Figure by the author.

challenges U.S. agriculture is facing. So not a Niagara Falls. Not even a leaky faucet.

Oh well, it's not like scientists apply for grant funding, right? Oh, wait. Yes, we do. In fact, we spend a good portion of our time—50 percent would not be an exaggeration—begging for money. Well, not begging but preparing grant applications: stating our hypotheses, describing our plans and their importance, calculating budgets, gathering letters of reference, and compiling bibliographies. But should we get lucky enough to get funded, administration takes their cut (anywhere from 10 to

100 percent of the incoming funds), and we struggle on to the next grant. We would not have jobs unless we brought in money from the outside. Don't believe me? Ask a scientist.

Maybe "pleading" would describe the process more accurately. But if politicians have their way, and funding is given only to researchers whose grants reinforce the politicians' beliefs, (good science!) "waste of time" would be a better description.

Chapter Fifteen

CO_2 IS PLANT FOOD

THE LAST BIT

"CO_2 IS *plant food!*"

"You keep saying that. I do not think it means what you think it means."

(My apologies to *The Princess Bride.*)

The bottom line, the last chance, the final fig leaf, is simple. As long as "climate change" and "CO_2 is plant food" remain political memes, research will stumble, and what research is done will be less than what is necessary. The public often see no need for research—the memes are political—a matter of opinion, of politics. "CO_2 is plant food" is a matter of belief, not fact.

Only when we can overcome the politics, when we can establish that this is not a question of viewpoint but of evidence, can we begin to address these issues. To respond, we need to understand the scope of the problem. Hence, the need for scientific research is immediate and pressing. But, catch-22, we can't do that because, hey, the problem is not really a problem, you see, just a difference of opinion, a matter of politics. A post-truth, an alternative fact.

There's more. For conservatives, the fundamental belief that government is bad, and hence government research is flawed, continues.[1] If you point out that every dollar spent on publicly funded agricultural research receives an annual rate of return of 35 to 100 percent to the economy,[2] they just won't believe it. (Jewish space lasers on the other hand . . .)

Yet a key reason that public funding of agricultural R&D generates such high returns is that it can create "spillovers," knowledge and tech that has wider applicability in other industries and regions. Such funding has been integral to the U.S. agricultural sector. And such research efforts have in part resulted in the global acknowledgement that the United States is the world's leading agricultural producer, the world's largest food exporter.[3]

One fundamental feature of being human and one reason for our success, or our failure, is our ability, or inability, to adapt as circumstances change. And science is key to adaptation. We begin an investigation to discover facts, good or bad, as they relate to human society. If we rely solely on the verbiage of "CO_2 is plant food," we avoid investigating the role that increasing CO_2 plays in plant biology.

As a former USDA scientist, I can state unequivocally that the research being done to understand and adapt to rising CO_2 is miniscule. The ARS may have about 1,200 scientists, but only a handful (fewer than five to my knowledge) are looking at CO_2 and plant biology. Unfortunately, many conservatives believe that science is not science but a political extension of liberalism.

This belief is creating a mental chasm between political will and scientific understanding that is reaching Grand Canyon proportions.

This view is anathema to every scientist I know. Science is capable of addressing questions related to CO_2 and plant biology. And in addressing these issues, there is hope. Hope that we can learn and adapt, that we can persevere when it comes to the challenge of rising CO_2 and climate shifts.

Are research and scientific discovery a panacea for all that ails humanity? No. But discovery, insight, and the innate human desire to explore, to understand how things work, and to make things better are fundamental to adaptation and, eventually, to solutions. Yet there are at present two paradigms when it comes to plant biology: one of "climate change," the other of "CO_2 is plant food." The idea that both could occur simultaneously, that one would affect the other, seems to be contrary to how issues of climate change are communicated to the public.

And there's the politics. Conservatives who insist that CO_2 is plant food and that if you were "green," you would support adding more of it to the atmosphere, global warming be damned, versus environmentalists who insist that any effect of CO_2 on plants will be inconsequential or wiped out by global warming.

But they are not separate issues. While I am happy to try and shed light on CO_2 effects per se in relation to plant biology, I also cannot ignore climate change. But unlike those who dismiss these impacts or regard them as negating anything positive about rising CO_2, I would argue that they can be exacerbating.

Does it matter? Yes. If you have made it this far, I hope I have convinced you that "CO_2 is plant food" matters a great deal. Is anything being done to integrate "CO_2 is plant food" research into the realm of scientific study needed to understand and respond to human-induced climate change?

Almost nothing. But the need to investigate further is not trivial; the links between plants and CO_2 are endemic to every aspect of civilization. Understanding those links and their consequences is essential to our survival and progress. To illustrate why this is, let me pose a handful of pragmatic questions. The last bits.

Given that CO_2 is plant food, and given the need to increase food production to account for an additional one to two billion individuals by midcentury, why aren't we pursuing a scientific program that will exploit additional CO_2 to increase seed yield?

It is clear from more than a hundred published studies that CO_2 will impact the nutritional quality of many staples like wheat and rice. Why aren't we documenting and quantifying these effects so that we can assess their impact on human health?

Why aren't we looking at CO_2-induced declines in plant protein for other animals in the food chain such as pollinators? Could the CO_2 effect be contributing to the "insect apocalypse"? What about other, more cuddly animals? What will happen to the pandas if bamboo leaves (their primary food source) become less nutritious?

It is clear that increasing CO_2 will differentially affect crops and weeds, with weeds generally showing a more favorable response. How will this impact future crop yields? Will it affect global food security?

How will the differential effect on crops and weeds alter pesticide use? Will we need to use more herbicide to control weeds? What are the economic consequences of that? What are the public health consequences?

For natural plant communities, we have seen preliminary evidence indicating that not all species in a forest respond in the same way to more CO_2. What impact will this have on biodiversity and long-term ecosystem function?

How will increasing CO_2 affect plant-based medicines? While synthetics are the primary source of medicines in developed countries, many people around the world depend on medicines made from plants. What impact will the CO_2 effect have on their availability?

If, as demonstrated earlier, CO_2 alters morphine production and potentially heroin production, how will this impact our ability to effectively wage our "war on drugs"? For example, if we were to reduce the acreage of illegal poppy production by 25 percent, would rising CO_2 increase production on the other 75 percent?

How will rising CO_2 affect the toxicology (poison content) of plants?

How will rising CO_2 interact with other climate change parameters, like extreme temperatures, drought, and flooding, to affect agriculture, pastures, prairies, forests, and wetlands?

If rising CO_2 can alter plant chemistry, what impact will it have on food-based allergies?

There are other pertinent questions to be asked, but you will have now gotten the gist of the level of our ignorance when it comes to "CO_2 is plant food." The fact remains that how CO_2 affects plant biology will also, like climate change, affect the world as a whole. I confess with all my being that I have no certified guarantee of answers as to what those effects might be. But I can offer thoughts—hypotheses, supposes, what ifs—that could change our thinking about "CO_2 is plant food." Here are an obvious few:

Face the meme. If anyone insists that "CO_2 is plant food," don't roll your eyes and ignore what you think is a silly statement. Confront it: "Really? Does that mean we will have more poison ivy in the future?"

Do the research. If I've done my job, you should have a pretty good idea of what more CO_2 would mean to plant biology, and if that isn't enough of an incentive, think about what it might mean to human health, from plant-based medicines to pollen allergies to nutrition to pesticide use.

Needless to say, research funding for such projects is nonexistent. Howard Frumkin is a Harvard-trained physician and epidemiologist who recently pointed out that out of a budget of $40 billion, the National Institutes of Health have devoted just $9 million to *all* climate change and public health research. That's 0.02 percent.[4] Would it kill us as a country if we increased that to 1 percent? And hey, maybe add a few dollars for "CO_2 is plant food" research, whose current NIH funding is a whopping 0 percent?

Let's go further. If you're the president of Nestlé, the world's largest food company, wouldn't it be prudent to do some research on CO_2 and its effect on, say, coffee? Or chocolate?

Of if you're Monsanto, the world's largest seed company, and you know that CO_2 can make plants grow, wouldn't you want to invest in varietal selection of wheat or rice or soybean or potato to find the variety that responds best to increasing CO_2? And then exploit that increased resource to make more seed? Obtain higher yields?

Or if you're Jeff Bezos, the world's richest man (with a net worth of about $170 billion as I write this), would it hurt you to spend 1 percent of your fortune trying to understand how rising CO_2 could affect the nutritional quality of staple foods for poor children throughout the world? Hell, how about 0.1 percent?

Seriously. Could we start living up to our "*sapiens*" billing?

Jibber jabber. When it comes to communicating science, scientists may not have the best skills for reaching nonscientists. Jargon and in-group speak are great for conferences but nonsensical to anyone else. And perhaps our inability to clearly communicate all the science behind climate change is a core issue. Sure, iconic images of drowning polar bears, rising sea levels, wildfires, and hurricanes resonate in the public square and are—correctly—associated with climate change. Clearly, plant scientists need to do a better job explaining the plant parts.

And could the media talk about this? And not just through the lens of "fossil fuels and CO_2 are wonderful"? Yes, I know it's complicated, but the rise of CO_2 represents a fundamental shift in all life as we know it. Isn't *that* worth a mention?

Chapter Sixteen

A PERSONAL NOTE

AT ONE TIME, science was a referee of political infighting. Something that could be agreed upon by both sides. I can recall the 2008 "We Can Solve It" global warming ad (sponsored by Al Gore's Alliance for Climate Protection) that starred—drumroll, please—Nancy Pelosi and Newt Gingrich![1] Since then, there has been a hardening of ideological lines when it comes to climate. Gingrich has, not surprisingly, refuted his participation in the ad ("Obviously, it was misconstrued")[2] and has gone on to call for the elimination of the Environmental Protection Agency altogether.

Ironically, while the science behind climate change has become more certain, the effort to discredit that science has grown more intense. There are a number of obvious reasons for this, most directly related to the bottom line: profits for fossil fuel interests. But there is one legal change that also deserves credit for the intensification: the 2010 Supreme Court ruling that "money is speech," also known as *Citizens United*.[3]

This ruling allowed individuals and groups to spend *unlimited* amounts of money on elections—with no accountability. To give you some sense of this from a fossil fuel–spending perspective, in 2006 the oil and gas industry spent $22 million on federal elections, in 2012 (the first post–*Citizens United* election) $73.5 million. 2016? $103 million. That is money that we know about—we do not know how much dark money was spent.[4] (Yes, $100 million dollars is a lot of money, but don't worry. Exxon's assets alone are about $350 billion. In percentage terms, an outlay of $100 million is just 0.028 percent of the company's net worth.)[5]

When this kind of money enters the political arena, science can no longer be an arbiter. When representatives who rely on oil and gas have unlimited funds, addressing climate change will not be high on the national agenda.

The fallout from *Citizens United* is not limited to climate science. If science interferes with your profits, and if those profits are stratospheric, then cash is slid to politicians to shut down any science that could attract attention or, worse yet, interfere, or—heavens!—impose regulations on your business. With enough cash, you can even spin in reverse: Having more guns is good! CO_2 is plant food!

But having unlimited, unaccountable funds to promote self-interest also promotes ignorance. When your goal is to cast aspersions on science, then science and its underlying tenets—experimentation, evidence, and facts—suffer. Objective truths, independent of beliefs, will not be revealed, and our knowledge of nature or chemistry or physics or medicine will fail.

The degree of censorship and political influence in denying science was unprecedented during the Trump administration. When objective truth is seen through an insecure political lens, fear arises. Fear that science will negate a political assertion and undermine a belief system that gave rise to that viewpoint. This is politics as religious fundamentalism, being scared of empirical truths.

Such fear was evident when Trump assumed office—and reflected in his efforts to degrade and demean climate science. He has called it a "hoax" on numerous occasions—and removed the United States from the Paris Climate Accords. We have now joined an elite group of countries that did not sign the agreement: Iran, Iraq, Libya, Eritrea, and Yemen.[6] And he did all this while simultaneously advocating for a seawall to protect his golf resort in Ireland from sea level rise.[7]

Trump's actions are not lone-wolf decisions; they are emblematic of a bought-and-paid-for Republican Party. In 2017, twenty-two Republican senators, including the then majority leader Mitch McConnell, signed a letter to Trump urging him to withdraw from the Paris Agreement. Forty conservative think tanks, including the Heritage Foundation, Grover Norquist's Americans for Tax Reform, the Koch family's Americans for Prosperity, and the Heartland Institute, signed a similar letter urging Trump to remove the United States from any climate change commitment.[8]

But climate change denial is only *one* aspect of the discord between politics and science when seen through a lens of fear. What happens when a pandemic like COVID-19 occurs? What are the options? Well, you can dismiss it ("We have it totally under control," January 22, 2020, "We pretty much shut it down coming in from China," February 2, 2020), offer nonscientific cures ("I see the disinfectant that knocks it out in a minute, one minute," April 23, 2020), or you can blame others ("This is nobody's fault but China," September 10, 2020).[9]

When reason is denied, when evidence is thrown out, when science is used a political means to an end, ignorance flourishes. And that ignorance has oozed its way into denying a pandemic, turning it into a political wedge issue.

What is particularly egregious in the politics-versus-science conflict is that science often doesn't fall into nice, well-defined categories. For example, is there a link between climate change and the COVID-19 pandemic? Constricted conservatives would pooh-pooh such a question. A scientist would be interested in finding out.

As I have stressed throughout this book, we depend on plants for food. And as humanity grows and as rising CO_2 threatens agriculture—either directly through weed competition or indirectly via climate change—our species will expand into new, untamed areas, squeezing wildlife—and the viruses they harbor—into the areas surrounding the space we occupy. Ebola, SARS, COVID-19. We are at the beginning of a new planetary dance with death. And science has never been more crucial to our survival.

Even without partisan denial, it is hard to face these threats. It's easier to overlook them, to push them aside. Ignorance by default. Yet the scale of the problems—the pandemics, climate, food, water, and biodiversity—is enormous. The scope is global. To ignore these issues for the sake of political expediency will continue to foster suffering and ensure the permanency of needless deaths. If there is a silver lining to the COVID-19 pandemic, it is the hope that individuals and future policy-makers will recognize what happens when political malfeasance joins hands with scientific ignorance—and take steps to prevent it.

One of the most fundamental steps is to reverse the political degradation of science. Let us elect leaders with the character, intelligence, and respect for logic, evidence, and reason necessary to face these problems. Reality is not a liberal hoax.

With denial of the COVID-19 pandemic on full, flagrant display, one can be forgiven for not recognizing the extent—the grossness—of Trump's efforts to pervert scientific inquiry.[10] Let me try to provide some insight into agricultural research and climate change during the Trump administration.

We can begin with Sonny Perdue, Secretary of Agriculture, who, as noted by *Politico*, has a questionable history of ethics violations, including signing a law giving himself a tax break and funding his campaign with contributions from his private enterprises. In any other administration, such ethics lapses would be glaring. But relative to other cabinet secretaries in the Trump administration, he falls into the choir boy category.[11]

Perdue is, of course, a skeptic of climate change, attributing it to "weather patterns."[12] Still, there's that pesky science, those facts that can throw shade on political assertions. What to do? It's hard to get rid of federal scientists.

Wait. What if we forced them to move? That might work. So Perdue hatched a new scheme, saying that the USDA's Economic Research Service (ERS) and the National Institute of Food and Agriculture (NIFA) could no longer reside in Washington,

DC, but had to move closer to agricultural lands. Kansas City, emblematic of the American agricultural heartland, was chosen. (Although I always think of KC as the best place for barbecue, but hey, it is closer to wheat fields.) Such a move is nonsensical, as the ERS deals with economics, and NIFA supplies research funding to land-grant universities. Economists and agricultural statisticians don't actually farm. Who knew?

Why these two agencies? Well, it turns out that the ERS had been disputing some of Trump's assertions on agriculture and the economy, specifically with regard to economic damage from climate change and food stamps. And NIFA had the temerity to publish a best-practices document for LGBTQ+ students in 4-H clubs, a youth program funded by the federal government.[13]

So forcing a move was in the political cards. Retribution. Not surprisingly, the experts, the folks at the ERS and NIFA, were less than thrilled. More than 50 percent of them resigned. Being experts, however, they did not, for the most part, join the unemployment rolls. They simply found jobs elsewhere, some for more money. Experts are valued, even government ones. Who knew?

But federal scientists were leaving! They were being kicked out for not following political mandates. Or as Mick Mulvaney, then Trump's chief of staff, bragged at a Republican fundraiser,

> I don't know if you saw the news the other day, but the USDA just moved two offices out of Washington, DC. . . . Guess what happened. Guess what happened. More than half the people quit. Now, it's nearly impossible to fire a federal worker. I know that because a lot of them work for me.

> And I've tried. . . . By simply saying to the people, "You know what, we're going to take you outside the bubble, outside the Beltway, outside this liberal haven [of Washington, DC] and move you out into the real part of the country," and they quit. What a wonderful way to streamline government and do what we haven't been able to do for a long time.[14]

The end result of all of this political Kabuki is not surprising. *Politico* obtained a mid-January 2020 USDA staffing memo stating that nearly two-thirds of the positions at the ERS were still unfilled five months after the move to Kansas City and that the ERS was having trouble with basic functions. Here is an excerpt from that memo, quoting an ERS scientist still employed at the agency[15]:

> One former ERS researcher described the situation as "a nightmare," adding that "it's quite clear that the capacity at the Economic Research Service has been diminished. I think there are many in the profession that don't expect the Economic Research Service to recover."

Are these agencies important? NIFA finances agricultural research fundamental to supporting and sustaining America's farmers, from finding ways to use pig manure to generate energy to developing drought-tolerant corn varieties. NIFA funds innovation and keeps U.S. agriculture globally competitive. As for the ERS, they look at economic trends in everything from organic agriculture to the fallout from *Salmonella*-related food poisoning.

Their science will be missed. American agriculture and you, the consumer, will suffer. Prices will rise, quality will diminish. I suspect that food security will become more of an issue in the next few years. I hope I am wrong.

Let me turn now to the Agricultural Research Service (ARS), the in-house research arm of the USDA. Its accomplishments are probably unknown to you, but they are incredibly diverse and include the mass production of penicillin during World War II (it turns out rotten cantaloupes were just the thing to encourage penicillin production), superabsorbent Pampers (corn starch), permanent-press slacks (wrinkle-resistant, easy-care cotton fabric), and saving Florida's citrus industry from citrus greening. It is a long list of achievements that have made our lives a little easier, American agriculture a little stronger, and food a little cheaper.

In essence, every dollar spent at the ARS generates about ten dollars in benefits to the economy.[16] But wait. Why can't private industry do this work? It can to some extent, but there is an important difference: Private R&D will invest in those products that are easy to patent and for which intellectual property rights can be obtained. All well and good, but other technologies, such as husbandry and management practices and efforts that require collective cooperation, like pest eradication, are not going to be profitable and thus require public investment.

Acknowledging climate change, developing appropriate adaptation and mitigation policies, improving water access and quality, fighting pests, promoting food safety, and ensuring nutrition: These are *public* investments, and one cannot assume that private industry will, out of the goodness of its heart, assume these responsibilities.

So what did the Trump administration do to hamstring the ARS? They couldn't move them, as it turns out that the ARS has laboratories around the country.[17] Hmm, let's see. Well, one of the most important positions in the USDA is chief scientist. As you can imagine, that is a role with a great deal of responsibility, and whoever occupies that position is, or should be, a scientist of remarkable caliber, someone who sets a scientific agenda.

Trump's first choice was Sam Clovis, a conservative radio host who had a PhD in public policy, but no science background, and was an ardent climate change denier. (He also questioned whether Barack Obama was born in the United States—surprise!). People who care about agriculture noticed. But it wasn't his lack of scientific credentials that did Mr. Clovis in. Rather, it was his ties to Russian interests. While serving on the Trump campaign, he supervised George Papadopoulos, who struck a plea deal on charges that he lied to FBI investigators about his communications with Russia-linked contacts. Clovis's withdrawal followed shortly after news of Papadopoulos's plea deal.[18]

The administration's second choice was Scott Hutchins, a seemingly more appropriate candidate as he had a PhD in entomology and had been a scientist for Dow Chemical. (Dow was a significant contributor to Trump's reelection campaign.) Perdue appointed Hutchins to the post. (To my knowledge, the appointment was never confirmed by the Senate.)

In looking at Hutchins's scientific background, however, I was struck by how few papers he had written. But one of those few caught my attention: "Natural Products for Crop

Protection: Evolution or Intelligent Design."[19] It concludes that evolution *and intelligent design* have combined to provide many important pest control products—many more are anticipated (my emphasis).

So the chief scientist of the USDA cannot make up his mind about evolution and believes in "intelligent design." But he was expected to be a leading advocate of scientific approaches to dealing with agriculture and climate change? In fairness, recent talks by Dr. Hutchins have at least shown that he is not a climate change denier. That's good. However, funding for the ARS tanked after his appointment. No resources were available to study climate change or CO_2 effects. Not so good.

As for me, I am proud to have worked at the ARS for twenty-four years. When I began in the mid-1990s, I worked in the Climate Stress Lab in Beltsville, Maryland, one of ten scientists looking at such issues as how crops were being affected by pollution, warmer temperatures, and increased ultraviolet radiation owing to ozone depletion. We were assessing not only impact but also evaluating and selecting for new crop varieties resistant to these stressors. But by 2018, I was the sole survivor. Everyone else had retired or passed on.

In May 2018, an article that I had worked on along with an international team of scientists from the Chinese Academy of Sciences, the University of Tokyo, the University of Washington, and the Bryan College of Health Sciences in Nebraska was about to be published in *Science Advances*, an offshoot of *Science*, a top-notch scientific journal.

The article was the culmination of two years of field trials throughout Asia evaluating the effect of rising CO_2 levels on the nutritional quality of rice. These were the first data on how CO_2 could alter the concentration of vitamins, including vitamins B and E, in rice (and the data I discussed in chapter 9).[20] Given that more than six hundred million people get 50 percent of their daily calories from rice, this was kind of a big deal. So as a courtesy, I called to let the USDA Office of Communications (OC) know that the paper was in press and would be coming out in a couple days.

A day later, I received an email from the OC stating that they could not issue a press release regarding the publication because the National Program Staff (NPS) did not agree with the conclusions of the paper. NPS thought that the paper was flawed, that the data did not support the conclusions.

This was odd for many reasons. For the paper even to be considered for publication, the NPS would have had to approve its submission. And the reasons given for rejecting the data didn't make any sense; they were based on a very superficial reading of the paper. Then there is the obvious: Administrators in general support the science of the agency they administrate. I responded to the concerns raised regarding the data and asked for a meeting to discuss them, but nothing happened.

Then another unusual action: The OC called up the University of Washington to urge them not to promote the paper. This was unprecedented—and dangerous. In part because the USDA gives grants to universities and colleges across the country, including the University of Washington. Calling up to try to influence someone not to do something where money is involved is closer to a mob tactic than a principle of higher education.

The University of Washington held firm. They not only approved the scientific validity of the paper but also promoted it.[21] And it got a lot of press.[22] Doing science is fine, but communicating it is equally important, especially when it's about a crop that affects hundreds of millions of people.

For me, it was a turning point. There was no reason to deny the results of a paper that had been peer-reviewed and was about to be published. It was unprecedented. And political. I was sixty-two years old and heavily invested after twenty-four years with the ARS. But I needed to leave. So I looked around and was extremely fortunate to be accepted by the Columbia University Mailman School of Public Health in 2019, where, thanks to Jeffrey Shaman's efforts (thanks, Jeff!), I can continue looking at the weird and wonderful nexus of climate, CO_2, plant biology, and public health, from poison ivy to nutrition, from pollen to parasites.

We fund science at the federal level to address questions of public concern for who and what we are as a nation. We have done so since our country was founded. Science is a cornerstone of democracy because it represents a means to derive truth. The only form of government that fears science is a dictatorship.

The damage inflicted by the Trump administration on scientific research was deep, a thousand bloody cuts. Now with the Biden administration—one that doesn't deny climate change—it is tempting to forget all that occurred, to renew and fight. But the impact of the delay in addressing climate science has been incalculable. And even now, the margin of those in political power who recognize the need to address science is razor-thin, subject to the illusory whims of those who deny the simplest of obvious mandates, like wearing a mask to stop the spread of an airborne disease. Even as the bleeding stops, one wonders if we can recover and address the climate monster knocking on the door. We are trying. Recently the Biden administration announced that the NIH can, maybe, if Congress approves, start to fund research on climate change and public health.[23] A budget of $100 million has been allocated. (That may seem like a lot, but it's a mere 0.015 percent of the national budget.)

I'm writing my research proposals now. Sigh.

CONCLUSION

When someone holds the door open for you, saying thank you is the thing to do—and you usually pass it on, holding the door open for the next person. For humans, small acts of kindness let us know that we are social animals. We want to help each other because life is difficult. We connect, we help. And when we give help, we receive it in return.

Doing research is holding open doors. It is looking at how the world works and looking for ways to make life better, a little easier for the person behind you. And there is no research area that helps others more than agriculture. More specifically, food. People need to eat, and going hungry is never fun. When hunger is extreme, it's life-threatening.

As a scientist, my greatest asset is truth. If people think that I am incapable of it, if they suppose that what I am telling them is tainted by bias, then they will not believe what I say or write. And those who have fought to deny, to disregard, to disparage climate science—the fossil fuel industry—have done their damnedest to blemish scientists who study climate change as promoting political bias.

They have done this in a campaign that has emulated tactics used by the tobacco industry to deny cause and effect, to deny the link between smoking and health consequences.[24] They recognize that science can be confusing for the public, so they simply fund a few well-placed contrarians in academia, and conflict ensues. Even if the vast majority of scientists who study climate change can point to the consequences, from CO_2 effects on nutrition to rising sea levels, all it takes—especially in a social media age—is a few well-placed deniers for science to become divided—and more to the point, for the public to become divided. After all, if scientists can't agree, why should we worry?

And in these efforts, they will take a single piece of data, wrap it in conjecture, and put it on a bumper sticker, all to deny climate science. There are a number of such denials, issued and repeated ad nauseum, but here are some of the most popular: Climate has changed before. The planet is cooling, not warming. It's the sun that's getting hotter. Scientists can't agree. Climate change is an academic hoax to get grant money. It hasn't warmed since 1998. And so on. If there were any truth in these claims, they would have been explored and documented in peer-reviewed scientific literature. But the experts, the climate scientists, have found no basis for these claims.

I would argue that the slipperiest of these claims is "CO_2 is plant food." Slippery because there is a scientific truth at the core of this assertion, and it takes a deeper dive to discover what it actually means in the context of plant biology—and climate change. I hope that what you have read here can be considered that deeper dive, giving you a better understanding of the consequences of that simple, deceitful meme, the rest of the story.

Scientists often believe that evidence (based on data) is beyond the political horizon and obvious to all. But science can be practiced and pursued only in a political framework that permits academic freedom. As the efforts of the Trump administration show, to ignore that scientific lens in the glare of a pandemic leads to ignorance—and the unnecessary deaths of many.

I am often asked about climate change, specifically "What can I do?" Yes, eating less meat, recycling, and using solar power are all useful, but the greatest impact you can have can be summed up in a single word:

Vote. Vote for science.

AFTERWORD AND THANKS

WRITING A book about science during a pandemic feels a lot like arguing whether the barn door should have been closed *before* the cows escaped. Pointing out the obvious is never fun.

But sometimes necessary. For all the words and images that have been used to describe the consequences of climate change, discussions of the role of CO_2 in plant biology—its fundamental importance and existential consequences—are met with silence and blank stares. Yet if scientists have been telling you that a pandemic was inevitable, and you were left dumbstruck and unprepared when one finally arrived, what will you do with the ongoing certainty of climate change and rising CO_2? Can you keep the barn door shut?

Will we succumb to our toxic blend of ignorance and arrogance? Will shiny dollars avert our gaze when it comes to the impacts on plant biology? Or will we learn? I hope for knowledge, for insight, for science. Ignorance is expensive, not only in terms of money but also in terms of character, vision, adaptation, potential, and who we—and future generations—aspire to be.

Such aspirations, such better angels, were kind enough to help me. To hold open a door. And I offer my thanks to these Angels for their time and wisdom.

Mark Bittman, famous chef and unparalleled advocate for food and health.

Greg Collister-Murray, friend and avid reader whose feedback lifted my spirits.

Allison Crimmins, EPA scientist who wears a cape and solves crime in her off-hours and who was recently tasked with leading the next National Climate Assessment. Brava!

Ruth DeFries, colleague with extraordinary knowledge of how the world works.

Carol Durst-Wertheim, food and science enthusiast and lover of life!

Mike Hoffman, a unique combination of Vietnam vet and Cornell entomologist.

Kim Knowlton, my eternal thanks for your time, kindness, encouragement.

Miranda Martin, for providing guideposts and discussion.

Marion Nestle, for her insight into the food chain and climate perspective.

Jeff Shaman, for taking a chance.

NOTES

PREFACE

1. Inconvenient Facts, "Inconvenient Facts 3 & 7 CO_2 Is Plant Food," YouTube video, 3:17, uploaded October 1, 2018, https://www.youtube.com/watch?v=wlzstC5zhkk.
2. Aaron Rupar, "Trump's Pick to Chair New Climate Panel Once Said CO_2 Has Been Maligned Like 'Jews Under Hitler,'" *Vox*, February 20, 2019, https://www.vox.com/2019/2/20/18233378/william-happer-trump-climate-change-panel.
3. Karin Kirk, "The Video Origin of the Myth that Global Warming Is Good for Agriculture," Yale Climate Connections, September 27, 2020, https://yaleclimateconnections.org/2020/09/video-origin-of-the-myth-that-global-warming-good-for-agriculture/.

1. PLANTS ARE IMPORTANT: THE PART ABOUT FOOD

1. Yinon M. Bar-On, Rob Phillips, and Ron Milo, "The Biomass Distribution on Earth," *Proceedings of the National Academy of Sciences* 115, no. 25 (2018): 6506–11.
2. Lifang Yang, Zhenyan Yang, Changkun Liu, Zhengshan He, Zhirong Zhang, Jing Yang, Haiyang Liu, Junbo Yang, and Yunheng Ji, "Chloroplast Phylogenomic Analysis Provides Insights into the Evolution of the

Largest Eukaryotic Genome Holder, *Paris japonica* (Melanthiaceae)," *BMC Plant Biology* 19, no. 1 (2019): 1–11.

3. Jared Diamond and Colin Renfrew, "Guns, Germs, and Steel: The Fates of Human Societies," *Nature* 386, no. 6623 (1997): 339–40; and S. Boyd Eaton, Stanley B. Eaton III, Melvin J. Konner, and Marjorie Shostak, "An Evolutionary Perspective Enhances Understanding of Human Nutritional Requirements," *Journal of Nutrition* 126, no. 6 (1996): 1732–40.
4. Jared Diamond, *Collapse: How Societies Choose to Fail or Succeed* (New York: Penguin, 2011).
5. Marc van de Mieroop, *A History of Ancient Egypt*, 2nd ed. (Hoboken, NJ: Wiley, 2021).
6. Roland Enmarch, ed., *The Dialogue of Ipuwer and the Lord of All* (Oxford: Griffith Institute, 2005).
7. Cormac Ó. Gráda, *Famine: A Short History* (Princeton, NJ: Princeton University Press, 2009).
8. Julian Cribb, *Food or War* (Cambridge: Cambridge University Press, 2019).
9. Dmitry Shaposhnikov, Boris Revich, Tom Bellander, Getahun Bero Bedada, Matteo Bottai, Tatyana Kharkova, Ekaterina Kvasha, et al., "Mortality Related to Air Pollution with the Moscow Heat Wave and Wildfire of 2010," *Epidemiology* 25, no. 3 (2014): 359.
10. Ines Perez, "Climate Change and Rising Food Prices Heightened Arab Spring," *Scientific American*, March 4, 2013, https://www.scientificamerican.com/article/climate-change-and-rising-food-prices-heightened-arab-spring/.
11. Colin P. Kelley, Shahrzad Mohtadi, Mark A. Cane, Richard Seager, and Yochanan Kushnir, "Climate Change in the Fertile Crescent and Implications of the Recent Syrian Drought," *Proceedings of the National Academy of Sciences* 112, no. 11 (2015): 3241–46.
12. John Podesta, *The Climate Crisis, Migration, and Refugees*, Brookings Institution, July 25, 2019, https://www.brookings.edu/research/the-climate-crisis-migration-and-refugees/.

2. PLANTS ARE IMPORTANT: THE PART ABOUT DRUGS

1. D. M. Qato, S. Zenk, J. Wilder, R. Harrington, D. Gaskin, and G. C. Alexander, "The Availability of Pharmacies in the United States: 2007–2015," *PLoS One* 12, no. 8 (2017): e0183172.

2. Florian P. Schiestl, "The Evolution of Floral Scent and Insect Chemical Communication," *Ecology Letters* 13, no. 5 (2010): 643–56.
3. "Coffee Statistics 2022," E-Imports, accessed February 28, 2022, https://www.e-importz.com/coffee-statistics.php#:~:text=Americans%20consume%20400%20million%20cups,with%20a%20great%20visible%20location.
4. Hannah Ritchie and Mark Roser, "Smoking," Our World in Data, May 2013 (updated January 2022), https://ourworldindata.org/smoking.
5. Marc Dhont, "History of Oral Contraception," *European Journal of Contraception and Reproductive Health Care* 15, no. S2 (2010): S12–18.
6. Gediya Shweta, Ribadiya Chetna, Soni Jinkal, Shah Nancy, and Jain Hitesh, "Herbal Plants Used as Contraceptives," *International Journal of Current Pharmaceutical Review and Research* 2, no. 1 (2011): 324–28.
7. J. Thomas Payte, "A Brief History of Methadone in the Treatment of Opioid Dependence: A Personal Perspective," *Journal of Psychoactive Drugs* 23, no. 2 (1991): 103–7.
8. "Ricin and the Umbrella Murder," CNN, October 23, 2003, https://www.cnn.com/2003/WORLD/europe/01/07/terror.poison.bulgarian/.
9. "Facts About Abrin," Centers for Disease Control and Prevention, accessed September 3, 2021, https://emergency.cdc.gov/agent/abrin/basics/facts.asp.
10. Boleslav L. Lichterman, "Aspirin: The Story of a Wonder Drug." *BMJ* 329, no. 7479 (2004): 1408.
11. "General Availability of Aspirin (100 mg) in the Public Health Sector," World Health Organization, accessed August 22, 2021, https://www.who.int/data/gho/data/indicators/indicator-details/GHO/general-availability-of-aspirin-(100-mg)-in-the-public-health-sector.
12. Centers for Disease Control and Prevention, *CDC National Health Report Highlights* (Atlanta: Centers for Disease Control and Prevention, 2014), https://www.cdc.gov/healthreport/publications/compendium.pdf.
13. Amber Dance, "As CBD Skyrockets in Popularity, Scientists Scramble to Understand How It's Metabolized," *Scientific American*, November 14, 2019, https://www.scientificamerican.com/article/as-cbd-skyrockets-in-popularity-scientists-scramble-to-understand-how-its-metabolized/#:~:text=%E2%80%9CIt%20seems%20that%20every%20corner,2024%2C%20according%20to%20one%20analysis.

3. PLANTS ARE IMPORTANT: THE PART ABOUT RELIGION

1. Alex S., "Henbane: Witch's Drug," Plant Profiles in Chemical Ecology: The Secret Life of Plants, January 30, 2017, https://sites.evergreen.edu/plantchemeco/henbane-medicine-andor-magic/.
2. Jean-Francois Sobiecki, "Psychoactive Ubulawu Spiritual Medicines and Healing Dynamics in the Initiation Process of Southern Bantu Diviners," *Journal of Psychoactive Drugs* 44, no. 3 (2012): 216–23.
3. Leander J. Valdés III, José Luis Díaz, and Ara G. Paul, "Ethnopharmacology of *ska María Pastora* (*Salvia divinorum*, Epling and Játiva-M.)," *Journal of Ethnopharmacology* 7, no. 3 (1983): 287–312.
4. Dennis J. McKenna, "Clinical Investigations of the Therapeutic Potential of Ayahuasca: Rationale and Regulatory Challenges," *Pharmacology and Therapeutics* 102, no. 2 (2004): 111–29.
5. Michael P. Bogenschutz and Alyssa A. Forcehimes, "Development of a Psychotherapeutic Model for Psilocybin-Assisted Treatment of Alcoholism," *Journal of Humanistic Psychology* 57, no. 4 (2017): 389–414.
6. Gus W. Van Beek, "Frankincense and Myrrh," *Biblical Archaeologist* 23, no. 3 (1960): 70–95.
7. David M. Watson, "Mistletoe: A Unique Constituent of Canopies Worldwide," *Forest Canopies* 2 (2004): 212–23.
8. "Meaning of Lotus Flower," One Tribe Apparel, October 24, 2018, https://www.onetribeapparel.com/blogs/pai/meaning-of-lotus-flower#:~:text=In%20Hinduism%2C%20the%20lotus%20flower,they%20represent%20purity%20and%20divinity.
9. William E. Ward, "The Lotus Symbol: Its Meaning in Buddhist Art and Philosophy," *Journal of Aesthetics and Art Criticism* 11, no. 2 (1952): 135–46.
10. Wikipedia, s.v. "Trees in Mythology," last modified January 3, 2022, https://en.wikipedia.org/wiki/Trees_in_mythology.
11. Robert Bevan-Jones, *The Ancient Yew: A History of* Taxus Baccata, 3rd ed. (Barnsley, UK: Windgather, 2016).
12. "Vasiliko (Basil Herb)," Greek Orthodox Christian Society, August/September 2016, https://lychnos.org/vasiliko-basil-herb/.
13. "Mahatma Gandhi Quotes," Brainy Quote, accessed March 8, 2022, https://www.brainyquote.com/authors/mahatma-gandhi-quotes.
14. "Quotes about God and Nature," Quote Master, accessed October 26, 2021, https://www.quotemaster.org/God+And+Nature.

4. PLANTS ARE IMPORTANT: THE PART ABOUT WEEDS

1. E.-C. Oerke, "Crop Losses to Pests," *Journal of Agricultural Science* 144, no. 1 (2006): 31–43.
2. Robert L. Zimdahl, *Fundamentals of Weed Science*, 5th ed. (Cambridge, MA: Academic, 2018).
3. Joseph M. DiTomaso, "Invasive Weeds in Rangelands: Species, Impacts, and Management," *Weed Science* 48, no. 2 (2000): 255–65.
4. James Howard Miller, *Nonnative Invasive Plants of Southern Forests: A Field Guide for Identification and Control*, General Technical Report SRS-62 (Asheville, NC: U.S. Department of Agriculture, Forest Service, Southern Research Station, 2003).
5. R. A. Wadsworth, Y. C. Collingham, S. G. Willis, B. Huntley, and P. E. Hulme, "Simulating the Spread and Management of Alien Riparian Weeds: Are They Out of Control?," *Journal of Applied Ecology* 37 (2000): 28–38.
6. L. G. Holm, L. W. Weldon, and R. D. Blackburn, "Aquatic Weeds," *Science* 166, no. 3906 (1969): 699–709.
7. F. L. Timmons, "A History of Weed Control in the United States and Canada," *Weed Science* 53, no. 6 (2005): 748–61.
8. "Herbicide," *Encyclopedia Britannica*, March 19, 2019, https://www.britannica.com/science/herbicide.
9. Stephen O. Duke, "The History and Current Status of Glyphosate," *Pest Management Science* 74, no. 5 (2018): 1027–34.
10. Stephen O. Duke and Stephen B. Powles, "Glyphosate: A Once-in-a-Century Herbicide," *Pest Management Science* 64, no. 4 (2008): 319–25; and Regina S. Baucom and Rodney Mauricio, "Fitness Costs and Benefits of Novel Herbicide Tolerance in a Noxious Weed," *Proceedings of the National Academy of Sciences* 101, no. 36 (2004): 13386–90.
11. Adam Liptak, "Supreme Court Supports Monsanto in Seed-Replication Case," *New York Times*, May 13, 2013, https://www.nytimes.com/2013/05/14/business/monsanto-victorious-in-genetic-seed-case.html.
12. "Estimated Annual Agricultural Pesticide Use, 2019," U.S. Geological Service, accessed October 25, 2021, https://water.usgs.gov/nawqa/pnsp/usage/maps/show_map.php?year=2019&map=GLYPHOSATE&hilo=L.
13. Ian Heap and Stephen O. Duke, "Overview of Glyphosate-Resistant Weeds Worldwide," *Pest Management Science* 74, no. 5 (2018): 1040–49.
14. Ian Heap, "Global Perspective of Herbicide-Resistant Weeds," *Pest Management Science* 70, no. 9 (2014): 1306–15.

15. Patricia Cohen, "Roundup Weedkiller Is Blamed for Cancers, but Farmers Say It's Not Going Away," *New York Times*, September 20, 2019, https://www.nytimes.com/2019/09/20/business/bayer-roundup.html.

5. PLANTS ARE IMPORTANT: THE PART ABOUT ART—AND ALLERGIES

1. Peg Herring, "The Secret Life of Soil," Oregon State University: OSU Extension Service, January 2010, https://extension.oregonstate.edu/news/secret-life-soil.
2. Kathryn Lasky, *The Most Beautiful Roof in the World: Exploring the Rainforest Canopy* (New York: Houghton Mifflin Harcourt, 2014).
3. Adam M. Lambert, Tom L. Dudley, and Kristin Saltonstall, "Ecology and Impacts of the Large-Statured Invasive Grasses *Arundo donax* and *Phragmites australis* in North America," *Invasive Plant Science and Management* 3, no. 4 (2010): 489–94.
4. Nicholas Outa, Dan Mungai, and James Last A. Keyombe, "The Impacts of Introduced Species on Lake Ecosystems: A Case of Lakes Victoria and Naivasha, Kenya," *AfricArXiv* (2019), https://doi.org/10.31730/osf.io/b5nyt.

6. SCIENCE IS FUNDAMENTAL

1. There is a great deal of information on global geological history, but some of the best explanations are available on Peter Hadfield's website. See potholer54, "5. Climate Change—Isn't It Natural?," YouTube video, 10:59, uploaded November 18, 2009, https://www.youtube.com/watch?v=w5hs4KVeiAU&list=PL82yk73N8eoX-Xobr_TfHsWPfAIyI7VAP&index=6&t=200s.
2. Yi Ge Zhang, Mark Pagani, Zhonghui Liu, Steven M. Bohaty, and Robert DeConto, "A 40-Million-Year History of Atmospheric CO_2," *Philosophical Transactions of the Royal Society A: Mathematical, Physical and Engineering Sciences* 371, no. 2001 (2013): 20130096.
3. Bruce A. Kimball, "Carbon Dioxide and Agricultural Yield: An Assemblage and Analysis of 430 Prior Observations," *Agronomy Journal* 75, no. 5 (1983): 779–88.
4. Robert Monroe, "Another Climate Milestone Falls at Mauna Loa Observatory," Scripps Institution of Oceanography, June 7, 2018, https://scripps.ucsd.edu/news/another-climate-milestone-falls-mauna-loa-observatory.

5. Arthur B. Robinson and Zachary W. Robinson, "Science Has Spoken: Global Warming Is a Myth," *Wall Street Journal*, December 4, 1997, https://www.wsj.com/articles/SB881189526293285000.

7. CO_2 IS PLANT FOOD: THE GOOD

1. L. H. Ziska and J. A. Bunce, "Predicting the Impact of Changing CO_2 on Crop Yields: Some Thoughts on Food," *New Phytologist* 175, no. 4 (2007): 607–18.
2. Herman Mayeux, Hyrum Johnson, Wayne Polley, and Stephen Malone, "Yield of Wheat Across a Subambient Carbon Dioxide Gradient," *Global Change Biology* 3, no. 3 (1997): 269–78.
3. Lewis H. Ziska, "Three-Year Field Evaluation of Early and Late 20th Century Spring Wheat Cultivars to Projected Increases in Atmospheric Carbon Dioxide," *Field Crops Research* 108, no. 1 (2008): 54–59.
4. Joe McCarthy, "2 Billion People Faced Food Insecurity Worldwide in 2019: UN Report," Global Citizen, July 15, 2020, https://www.globalcitizen.org/en/content/state-of-food-security-and-nutrition-report-2020/#:~:text=An%20estimated%20746%20million%20people,World%202020%20report%20released%20Tuesday.
5. "Food Security and COVID-19," World Bank, last updated December 13, 2021, https://www.worldbank.org/en/topic/agriculture/brief/food-security-and-covid-19.
6. Lewis H. Ziska, Paz A. Manalo, and Raymond A. Ordonez, "Intraspecific Variation in the Response of Rice (*Oryza sativa* L.) to Increased CO_2 and Temperature: Growth and Yield Response of 17 Cultivars," *Journal of Experimental Botany* 47, no. 9 (1996): 1353–59.
7. Lewis H. Ziska, James A. Bunce, Hiroyuki Shimono, David R. Gealy, Jeffrey T. Baker, Paul C. D. Newton, Matthew P. Reynolds, et al., "Food Security and Climate Change: On the Potential to Adapt Global Crop Production by Active Selection to Rising Atmospheric Carbon Dioxide," *Proceedings of the Royal Society B: Biological Sciences* 279, no. 1745 (2012): 4097–105.
8. Lewis H. Ziska and Anna McClung, "Differential Response of Cultivated and Weedy (Red) Rice to Recent and Projected Increases in Atmospheric Carbon Dioxide," *Agronomy Journal* 100, no. 5 (2008): 1259–63.

9. A. Raschi, F. Miglietta, R. Tognetti, and P. R. van Gardingen, eds., *Plant Responses to Elevated CO_2: Evidence from Natural Springs* (Cambridge: Cambridge University Press, 1997).
10. Paul C. D. Newton, R. Andrew Carran, Grant R. Edwards, and Pascal A. Niklaus, eds., *Agroecosystems in a Changing Climate* (Boca Raton, FL: CRC, 2006).
11. L. H. Ziska and D. M. Blumenthal, "Empirical Selection of Cultivated Oat in Response to Rising Atmospheric Carbon Dioxide," *Crop Science* 47, no. 4 (2007): 1547–52.
12. L. H. Ziska, C. F. Morris, and E. W. Goins, "Quantitative and Qualitative Evaluation of Selected Wheat Varieties Released Since 1903 to Increasing Atmospheric Carbon Dioxide: Can Yield Sensitivity to Carbon Dioxide Be a Factor in Wheat Performance?," *Global Change Biology* 10, no. 10 (2004): 1810–19.
13. Hidemitsu Sakai, Takeshi Tokida, Yasuhiro Usui, Hirofumi Nakamura, and Toshihiro Hasegawa, "Yield Responses to Elevated CO_2 Concentration Among Japanese Rice Cultivars Released Since 1882," *Plant Production Science* 22, no. 3 (2019): 352–66.
14. X.-G. Zhu, A. R. Portis Jr., and S. P. Long, "Would Transformation of C3 Crop Plants with Foreign Rubisco Increase Productivity? A Computational Analysis Extrapolating from Kinetic Properties to Canopy Photosynthesis," *Plant, Cell & Environment* 27, no. 2 (2004): 155–65.
15. "Russet Burbank," Wikipedia, last modified December 15, 2021, https://en.wikipedia.org/wiki/Russet_Burbank.
16. "International Rice Genebank," International Rice Research Institute, accessed October 26, 2021, https://www.irri.org/international-rice-genebank.
17. The DOE is often referred to as the "Department of Everything" because of its large budget.
18. George R. Hendrey, David S. Ellsworth, Keith F. Lewin, and John Nagy, "A Free-Air Enrichment System for Exposing Tall Forest Vegetation to Elevated Atmospheric CO_2," *Global Change Biology* 5, no. 3 (1999): 293–309.
19. Joseph A. M. Holtum and Klaus Winter, "Photosynthetic CO_2 Uptake in Seedlings of Two Tropical Tree Species Exposed to Oscillating Elevated Concentrations of CO_2," *Planta* 218, no. 1 (2003): 152–58; and L. H. Allen, B. A. Kimball, J. A. Bunce, M. Yoshimoto, Y. Harazono, J. T. Baker, K. J. Boote, and J. W. White, "Fluctuations of CO_2 in Free-Air CO_2 Enrichment (FACE) Depress Plant Photosynthesis, Growth, and Yield," *Agricultural and Forest Meteorology* 284 (2020): 107899.

20. "CO_2 Gas Enrichment for Crops," AG Gas, accessed October 25, 2021, https://www.carbogation.com/.
21. John P. Renschler, Kelsey M. Walters, Paul N. Newton, and Ramanan Laxminarayan, "Estimated Under-Five Deaths Associated with Poor-Quality Antimalarials in Sub-Saharan Africa," *American Journal of Tropical Medicine and Hygiene* 92, no. S6 (2015): 119–26.
22. Sanne de Ridder, Frank van der Kooy, and Robert Verpoorte, "*Artemisia annua* as a Self-Reliant Treatment for Malaria in Developing Countries," *Journal of Ethnopharmacology* 120, no. 3 (2008): 302–14.
23. "Treating Malaria," World Health Organization, accessed October 25, 2021, https://www.who.int/malaria/areas/treatment/overview/en/#:~:text=WHO%20recommends%20artemisinin%2Dbased%20combination,effective%20antimalarial%20medicines%20available%20today.
24. C. Zhu, Q. Zeng, A. McMichael, K. L. Ebi, K. Ni, A. S. Khan, J. Zhu, et al., "Historical and Experimental Evidence for Enhanced Concentration of Artemesinin, a Global Anti-malarial Treatment, with Recent and Projected Increases in Atmospheric Carbon Dioxide," *Climatic Change* 132 (2015): 295–306.

8. CO_2 IS PLANT FOOD: THE BAD

1. Colin Khoury, "How Many Plants Feed the World?," *Agricultural Biodiversity Weblog*, June 8, 2010, https://agro.biodiver.se/2010/06/how-many-plants-feed-the-world/.
2. David C. Bridges, *Crop Losses Due to Weeds in the United States, 1992* (Westminster, CO: Weed Science Society of America, 1992).
3. Roy J. Smith, "Weed Thresholds in Southern US Rice, *Oryza sativa*," *Weed Technology* 2, no. 3 (1988): 232–41.
4. L. H. Ziska, M. B. Tomecek, and D. R. Gealy, "Competitive Interactions Between Cultivated and Red Rice as a Function of Recent and Projected Increases in Atmospheric Carbon Dioxide," *Agronomy Journal* 102, no. 1 (2010): 118–23.
5. Lewis H. Ziska, "Global Climate Change and Carbon Dioxide: Assessing Weed Biology and Management," in *Handbook on Climate Change and Agroecosystems: Impacts, Adaptation, and Mitigation*, ed. Daniel Hillel and Cynthia Rosenzweig (London: Imperial College, 2011), 191–208.

6. Jonas Vengris, "Plant Nutrient Competition Between Weeds and Corn," *Agronomy Journal* 47 (1955): 213–15.
7. G. J. A. Ryle and J. Stanley, "Effect of Elevated CO_2 on Stomatal Size and Distribution in Perennial Ryegrass," *Annals of Botany* 69, no. 6 (1992): 563–65.
8. Mithila Jugulam, Aruna K. Varanasi, Vijaya K. Varanasi, and P. V. V. Prasad, "Climate Change Influence on Herbicide Efficacy and Weed Management," in *Food Security and Climate Change*, ed. Shyam S. Yadav, Robert J. Redden, Jerry L. Hatfield, Andreas W. Ebert, and Danny Hunter (Hoboken, NJ: Wiley, 2018), 433–48.
9. Pawel Waryszak, Tanja I. Lenz, Michelle R. Leishman, and Paul O. Downey, "Herbicide Effectiveness in Controlling Invasive Plants Under Elevated CO_2: Sufficient Evidence to Rethink Weeds Management," *Journal of Environmental Management* 226 (2018): 400–407.
10. Kerri Skinner, Lincoln Smith, and Peter Rice, "Using Noxious Weed Lists to Prioritize Targets for Developing Weed Management Strategies," *Weed Science* 48, no. 5 (2000): 640–44.
11. Lewis H. Ziska, Shaun Faulkner, and John Lydon, "Changes in biomass and root: Shoot Ratio Of Field-Grown Canada Thistle (*Cirsium arvense*), a Noxious, Invasive Weed, with Elevated CO_2: Implications for Control with Glyphosate," *Weed Science* 52, no. 4 (2004): 584–88; and Lewis H. Ziska, "Elevated Carbon Dioxide Alters Chemical Management of Canada Thistle in No-Till Soybean," *Field Crops Research* 119, nos. 2–3 (2010): 299–303.
12. Jose V. Tarazona, Manuela Tiramani, Hermine Reich, Rudolf Pfeil, Frederique Istace, and Federica Crivellente, "Glyphosate Toxicity and Carcinogenicity: A Review of the Scientific Basis of the European Union Assessment and Its Differences with IARC," *Archives of Toxicology* 91, no. 8 (2017): 2723–43.
13. Lewis H. Ziska, Martha B. Tomecek, and David R. Gealy, "Competitive Interactions Between Cultivated and Red Rice as a Function of Recent and Projected Increases in Atmospheric Carbon Dioxide," *Agronomy Journal* 102, no. 1 (2010): 118–23.
14. Lewis H. Ziska, David R. Gealy, Martha B. Tomecek, Aaron K. Jackson, and Howard L. Black, "Recent and Projected Increases in Atmospheric CO_2 Concentration Can Enhance Gene Flow Between Wild and Genetically Altered Rice (*Oryza sativa*)," *PLoS One* 7, no. 5 (2012): e37522.

15. João Paulo Refatti, Luis Antonio de Avila, Edinalvo Rabaioli Camargo, Lewis Hans Ziska, Claudia Oliveira, Reiofeli Salas-Perez, Christopher Edward Rouse, and Nilda Roma-Burgos, "High [CO_2] and Temperature Increase Resistance to Cyhalofop-butyl in Multiple-Resistant *Echinochloa colona*," *Frontiers in Plant Science* 10 (2019): 529.
16. Refatti et al., "High [CO_2] and Temperature Increase Resistance to Cyhalofop-butyl."
17. "Experiment Suggests Limitation to CO_2 Tree Banking," *Duke Today*, August 7, 2007, https://today.duke.edu/2007/08/carbonadd.html.
18. Jacqueline E. Mohan, Lewis H. Ziska, William H. Schlesinger, Richard B. Thomas, Richard C. Sicher, Kate George, and James S. Clark, "Biomass and Toxicity Responses of Poison Ivy (*Toxicodendron radicans*) to Elevated Atmospheric CO_2," *Proceedings of the National Academy of Sciences* 103, no. 24 (2006): 9086–89.
19. "Cheatgrass, Downy Brome," EDDMapS, accessed October 26, 2021, https://www.eddmaps.org/distribution/usstate.cfm?sub=5214.
20. Bethany A. Bradley, Caroline A. Curtis, Emily J. Fusco, John T. Abatzoglou, Jennifer K. Balch, Sepideh Dadashi, and Mao-Ning Tuanmu, "Cheatgrass (*Bromus tectorum*) Distribution in the Intermountain Western United States and Its Relationship to Fire Frequency, Seasonality, and Ignitions," *Biological Invasions* 20, no. 6 (2018): 1493–1506.
21. Hal Bernton, "'Grassoline' Fueling the Spread of Washington's Wildfires," *Seattle Times*, August 22, 2015, https://www.seattletimes.com/seattle-news/environment/pesky-pervasive-cheatgrass-feeds-regions-fires-2/.
22. L. H. Ziska, J. B. Reeves III, and B. Blank, "The Impact of Recent Increases in Atmospheric CO_2 on Biomass Production and Vegetative Retention of Cheatgrass (*Bromus tectorum*): Implications for Fire Disturbance," *Global Change Biology* 11, no. 8 (2005): 1325–32; and Robert R. Blank, Tye Morgan, Lewis H. Ziska, and Robert H. White, "Effect of Atmospheric CO_2 Levels on Nutrients in Cheatgrass Tissue," *Natural Resources and Environmental Issues* 16, no. 1 (2011): 18.
23. Mike Pellant and Christi Hall, "Distribution of Two Exotic Grasses on Intermountain Rangelands: Status in 1992," in *Proceedings in Ecology and Management of Annual Rangelands*, General Technical Report INT-GTR-313, ed. Stephen B. Monsen and Stanley G. Ketchum (Ogden, UT: U.S. Department of Agriculture, Forest Service, Intermountain Research Station, 1994), 109–12; and Mike Pellant, *Cheatgrass: The Invader That*

Won the West (Boise, ID: U.S. Department of the Interior, Bureau of Land Management, Idaho State Office, 1996).

24. Stanley D. Smith, Travis E. Huxman, Stephen F. Zitzer, Therese N. Charlet, David C. Housman, James S. Coleman, Lynn K. Fenstermaker, Jeffrey R. Seemann, and Robert S. Nowak, "Elevated CO_2 Increases Productivity and Invasive Species Success in an Arid Ecosystem," *Nature* 408, no. 6808 (2000): 79–82.
25. Dana M. Blumenthal, Víctor Resco, Jack A. Morgan, David G. Williams, Daniel R. LeCain, Erik M. Hardy, Elise Pendall, and Emma Bladyka, "Invasive Forb Benefits from Water Savings by Native Plants and Carbon Fertilization Under Elevated CO_2 and Warming," *New Phytologist* 200, no. 4 (2013): 1156–65.
26. Blumenthal et al., "Invasive Forb Benefits."
27. S. D. Smith, B. R. Strain, and T. D. Sharkey, "Effects of CO_2 Enrichment on Four Great Basin Grasses," *Functional Ecology* (1987): 139–43.
28. Christopher R. Webster, Michael A. Jenkins, and Shibu Jose, "Woody Invaders and the Challenges They Pose to Forest Ecosystems in the Eastern United States," *Journal of Forestry* 104, no. 7 (2006): 366–74.
29. Harold A. Mooney and Richard J. Hobbs, "Global Change and Invasive Species: Where Do We Go from Here?," in *Invasive Species in a Changing World*, ed. Harold A. Mooney and Richard J. Hobbs (Washington, DC: Island, 2000), 425–34.
30. Liba Pejchar and Harold A. Mooney, "Invasive Species, Ecosystem Services and Human Well-Being," *Trends in Ecology & Evolution* 24, no. 9 (2009): 497–504.
31. Oliver L. Phillips, Rodolfo Vásquez Martínez, Luzmila Arroyo, Timothy R. Baker, Timothy Killeen, Simon L. Lewis, Yadvinder Malhi, et al., "Increasing Dominance of Large Lianas in Amazonian Forests," *Nature* 418, no. 6899 (2002): 770–74.

9. THE OMG

1. "Nutrient Composition and Protein Quality of Rice Relative to Other Cereals," Food and Agricultural Organization of the United Nations, accessed October 26, 2021, http://www.fao.org/3/t0567e/T0567E0d.htm.
2. Jann P. Conroy, "Influence of Elevated Atmospheric CO_2 Concentrations on Plant Nutrition," *Australian Journal of Botany* 40, no. 5 (1992): 445–56.

3. Alina Petre, "21 Vegetarian Foods That Are Loaded with Iron," *Healthline*, May 4, 2017, https://www.healthline.com/nutrition/iron-rich-plant-foods.
4. Ananda S. Prasad, *Zinc in Human Nutrition* (Boca Raton, FL: CRC, 1979).
5. Chunwu Zhu, Kazuhiko Kobayashi, Irakli Loladze, Jianguo Zhu, Qian Jiang, Xi Xu, Gang Liu, et al., "Carbon Dioxide (CO_2) Levels This Century Will Alter the Protein, Micronutrients, and Vitamin Content of Rice Grains with Potential Health Consequences for the Poorest Rice-Dependent Countries," *Science Advances* 4, no. 5 (2018): eaaq1012.
6. "Essential Food for the Poor," *Rice Today*, September 9, 2002, https://ricetoday.irri.org/essential-food-for-the-poor/.
7. Irakli Loladze, "Rising Atmospheric CO_2 and Human Nutrition: Toward Globally Imbalanced Plant Stoichiometry?," *Trends in Ecology & Evolution* 17, no. 10 (2002): 457–61.
8. Samuel S. Myers, Antonella Zanobetti, Itai Kloog, Peter Huybers, Andrew D. B. Leakey, Arnold J. Bloom, Eli Carlisle, et al., "Increasing CO_2 Threatens Human Nutrition," *Nature* 510, no. 7503 (2014): 139–42.
9. Robert H. Beach, Timothy B. Sulser, Allison Crimmins, Nicola Cenacchi, Jefferson Cole, Naomi K. Fukagawa, Daniel Mason-D'Croz, et al., "Combining the Effects of Increased Atmospheric Carbon Dioxide on Protein, Iron, and Zinc Availability and Projected Climate Change on Global Diets: A Modelling Study," *Lancet Planetary Health* 3, no. 7 (2019): e307–17.
10. Daniel R. Taub, Brian Miller, and Holly Allen, "Effects of Elevated CO_2 on the Protein Concentration of Food Crops: A Meta-analysis," *Global Change Biology* 14, no. 3 (2008): 565–75.
11. Donald R. Davis, "Trade-Offs in Agriculture and Nutrition," *Food Technology* 59, no. 3 (2005): 120.
12. Lewis H. Ziska, Jeffery S. Pettis, Joan Edwards, Jillian E. Hancock, Martha B. Tomecek, Andrew Clark, Jeffrey S. Dukes, Irakli Loladze, and H. Wayne Polley, "Rising Atmospheric CO_2 is Reducing the Protein Concentration of a Floral Pollen Source Essential for North American Bees," *Proceedings of the Royal Society B: Biological Sciences* 283, no. 1828 (2016): 20160414.
13. Ziska et al., "Rising Atmospheric CO_2 is Reducing the Protein Concentration."
14. T'ai H. Roulston, James H. Cane, and Stephen L. Buchmann, "What Governs Protein Content of Pollen: Pollinator Preferences, Pollen–Pistil

Interactions, or Phylogeny?," *Ecological Monographs* 70, no. 4 (2000): 617–43.

15. David J. Augustine, Dana M. Blumenthal, Tim L. Springer, Daniel R. LeCain, Stacey A. Gunter, and Justin D. Derner, "Elevated CO_2 Induces Substantial and Persistent Declines in Forage Quality Irrespective of Warming in Mixedgrass Prairie," *Ecological Applications* 28, no. 3 (2018): 721–35; and Ellen A. R. Welti, Karl A. Roeder, Kirsten M. de Beurs, Anthony Joern, and Michael Kaspari, "Nutrient Dilution and Climate Cycles Underlie Declines in a Dominant Insect Herbivore," *Proceedings of the National Academy of Sciences* 117, no. 13 (2020): 7271–75.
16. Becky Upham, "Poor Nutrition in the U.S. Poses Threats to Health, National Security, and Economy, Panel Says," *Everyday Health*, July 28, 2020, https://www.everydayhealth.com/diet-nutrition/poor-nutrition-in-us-poses-threats-to-health-national-security-and-economy-panel-says/.
17. Charles W. Schmidt, "Pollen Overload: Seasonal Allergies in a Changing Climate," *Environmental Health Perspectives* 124, no. 4 (2016): A70–75.
18. Michael Kerr, "Pollen Library: Plants That Cause Allergies," *Health Line*, November 13, 2018, https://www.healthline.com/health/allergies/pollen-library#1.
19. Lewis H. Ziska and Frances A. Caulfield, "Rising CO_2 and Pollen Production of Common Ragweed (*Ambrosia artemisiifolia* L.), a Known Allergy-Inducing Species: Implications for Public Health," *Functional Plant Biology* 27, no. 10 (2000): 893–98.
20. Ben D. Singer, Lewis H. Ziska, David A. Frenz, Dennis E. Gebhard, and James G. Straka, "Research Note: Increasing Amb a 1 Content in Common Ragweed (*Ambrosia artemisiifolia*) Pollen as a Function of Rising Atmospheric CO_2 Concentration," *Functional Plant Biology* 32, no. 7 (2005): 667–70.
21. Kyu Rang Kim, Jae-Won Oh, Su-Young Woo, Yun Am Seo, Young-Jin Choi, Hyun Seok Kim, Wi Young Lee, and Baek-Jo Kim, "Does the Increase in Ambient CO_2 Concentration Elevate Allergy Risks Posed by Oak Pollen?," *International Journal of Biometeorology* 62, no. 9 (2018): 1587–94.
22. Bert Brunekreef, Gerard Hoek, Paul Fischer, and Frits Th M. Spieksma, "Relation Between Airborne Pollen Concentrations and Daily Cardiovascular and Respiratory-Disease Mortality," *Lancet* 355, no. 9214 (2000): 1517–18.

23. Lewis H. Ziska, László Makra, Susan K. Harry, Nicolas Bruffaerts, Marijke Hendrickx, Frances Coates, Annika Saarto, et al., "Temperature-Related Changes in Airborne Allergenic Pollen Abundance and Seasonality Across the Northern Hemisphere: A Retrospective Data Analysis," *Lancet Planetary Health* 3, no. 3 (2019): e124–31.
24. Ruchi S. Gupta, Christopher M. Warren, Bridget M. Smith, Jesse A. Blumenstock, Jialing Jiang, Matthew M. Davis, and Kari C. Nadeau, "The Public Health Impact of Parent-Reported Childhood Food Allergies in the United States," *Pediatrics* 142, no. 6 (2018): e20181235.
25. M. Bannayan, C. M. Tojo Soler, L. C. Guerra, and G. Hoogenboom, "Interactive Effects of Elevated [CO_2] and Temperature on Growth and Development of a Short- and Long-Season Peanut Cultivar," *Climatic Change* 93, no. 3 (2009): 389–406.
26. Lewis H. Ziska, Jinyoung Yang, Martha B. Tomecek, and Paul J. Beggs, "Cultivar-Specific Changes in Peanut Yield, Biomass, and Allergenicity in Response to Elevated Atmospheric Carbon Dioxide Concentration," *Crop Science* 56, no. 5 (2016): 2766–74.
27. Hipolito Nzwalo and Julie Cliff, "Konzo: From Poverty, Cassava, and Cyanogen Intake to Toxico-nutritional Neurological Disease," *PLoS Neglected Tropical Diseases* 5, no. 6 (2011): e1051.
28. Roslyn M. Gleadow, John R. Evans, Stephanie McCaffery, and Timothy R. Cavagnaro, "Growth and Nutritive Value of Cassava (*Manihot esculenta* Cranz.) Are Reduced When Grown in Elevated CO_2," *Plant Biology* 11 (2009): 76–82.
29. Roslyn M. Gleadow and Birger Lindberg Møller, "Cyanogenic Glycosides: Synthesis, Physiology, and Phenotypic Plasticity," *Annual Review of Plant Biology* 65 (2014): 155–85.
30. Daniel A. Potter and David W. Held, "Biology and Management of the Japanese Beetle," *Annual Review of Entomology* 47, no. 1 (2002): 175–205.
31. Jorge A. Zavala, Clare L. Casteel, Evan H. DeLucia, and May R. Berenbaum, "Anthropogenic Increase in Carbon Dioxide Compromises Plant Defense Against Invasive Insects," *Proceedings of the National Academy of Sciences* 105, no. 13 (2008): 5129–33.
32. L. H. Ziska, S. D. Emche, E. L. Johnson, K. A. T. E. George, D. R. Reed, and R. C. Sicher, "Alterations in the Production and Concentration of Selected Alkaloids as a Function of Rising Atmospheric Carbon Dioxide and Air Temperature: Implications for Ethno-pharmacology," *Global Change Biology* 11, no. 10 (2005): 1798–1807.

33. Xin Li, Lan Zhang, Golam Jalal Ahammed, Zhi-Xin Li, Ji-Peng Wei, Chen Shen, Peng Yan, Li-Ping Zhang, and Wen-Yan Han, "Stimulation in Primary and Secondary Metabolism by Elevated Carbon Dioxide Alters Green Tea Quality in *Camellia sinensis* L.," *Scientific Reports* 7, no. 1 (2017): 1–12.
34. Fernando E. Vega, Lewis H. Ziska, Ann Simpkins, Francisco Infante, Aaron P. Davis, Joseph A. Rivera, Jinyoung Y. Barnaby, and Julie Wolf, "Early Growth Phase and Caffeine Content Response to Recent and Projected Increases in Atmospheric Carbon Dioxide in Coffee (*Coffea arabica* and *C. canephora*)," *Scientific Reports* 10, no. 1 (2020): 1–11.
35. Drug Enforcement Administration Museum (website), accessed October 26, 2021, https://www.deamuseum.org/ccp/opium/history.html.
36. Eric Palmer, "Sun Gets Direct Line to Poppy Production with Buyout of GSK Opiates Biz," *Fierce Pharma*, September 3, 2015, https://www.fiercepharma.com/supply-chain/sun-gets-direct-line-to-poppy-production-buyout-of-gsk-opiates-biz.
37. Anne S. van Wyk and Gerhard Prinsloo, "Health, Safety and Quality Concerns of Plant-Based Traditional Medicines and Herbal Remedies," *South African Journal of Botany* 133 (2020): 54–62.
38. Lewis H. Ziska, Sini Panicker, and Heidi L. Wojno, "Recent and Projected Increases in Atmospheric Carbon Dioxide and the Potential Impacts on Growth and Alkaloid Production in Wild Poppy (*Papaver setigerum* DC.)," *Climatic Change* 91, no. 3 (2008): 395–403.
39. Ziska et al., "Recent and Projected Increases in Atmospheric Carbon Dioxide."
40. Dawn Connelly, "A History of Aspirin," *Pharmaceutical Journal*, September 26, 2014, https://www.pharmaceutical-journal.com/news-and-analysis/infographics/a-history-of-aspirin/20066661.article?firstPass=false.
41. "Aspirin," Wikipedia, last modified January 3, 2022, https://en.wikipedia.org/wiki/Aspirin.
42. S. Hosztafi, "The History of Heroin," *Acta Pharmaceutica Hungarica* 71, no. 2 (2001): 233–42.
43. "Provisional Accounts of Drug Overdose Deaths, as of 8/6/2017," Centers for Disease Control and Prevention, accessed October 26, 2021, https://www.cdc.gov/nchs/data/health_policy/monthly-drug-overdose-death-estimates.pdf.
44. "A Bright Future for Loblolly Pine Forests," CO_2 Coalition, December 13, 2018, https://co2coalition.org/2018/12/13/a-bright-future-for-u-s-loblolly-pine-forests/.

45. Aaron Rupar, "Trump's Pick to Chair New Climate Panel Once Said CO_2 Has Been Maligned Like 'Jews Under Hitler,'" *Vox*, February 20, 2019, https://www.vox.com/2019/2/20/18233378/william-happer-trump-climate-change-panel.
46. William F. Laurance, Alexandre A. Oliveira, Susan G. Laurance, Richard Condit, Christopher W. Dick, Ana Andrade, Henrique EM Nascimento, Thomas E. Lovejoy, and José ELS Ribeiro, "Altered Tree Communities in Undisturbed Amazonian Forests: A Consequence of Global Change?," *Biotropica* 37, no. 2 (2005): 160–62.
47. Irwin N. Forseth and Anne F. Innis, "Kudzu (*Pueraria montana*): History, Physiology, and Ecology Combine to Make a Major Ecosystem Threat," *Critical Reviews in Plant Sciences* 23, no. 5 (2004): 401–13.
48. Amy E. Wiberley, Autumn R. Linskey, Tanya G. Falbel, and Thomas D. Sharkey, "Development of the Capacity for Isoprene Emission in Kudzu," *Plant, Cell & Environment* 28, no. 7 (2005): 898–905.
49. Jake F. Weltzin, R. Travis Belote, and Nathan J. Sanders, "Biological Invaders in a Greenhouse World: Will Elevated CO_2 Fuel Plant Invasions?," *Frontiers in Ecology and the Environment* 1, no. 3 (2003): 146–53; and Thomas W. Sasek and Boyd R. Strain, "Effects of Carbon Dioxide Enrichment on the Expansion and Size of Kudzu (*Pueraria lobata*) Leaves," *Weed Science* 37, no. 1 (1989): 23–28.
50. "Wildfire Smoke and COVID-19: Frequently Asked Questions and Resources for Air Resource Advisors and Other Environmental Health Professionals," Centers for Disease Control and Prevention, last updated October 9, 2020, https://www.cdc.gov/coronavirus/2019-ncov/php/smoke-faq.html.

PART III. CO_2 IS PLANT FOOD. NOW WHAT?

1. "AFRI Request for Applications," National Institute for Food and Agriculture, U.S. Department of Agriculture, accessed October 26, 2021, https://nifa.usda.gov/afri-request-applications.

10. MORE QUESTIONS THAN ANSWERS

1. Kevin J. Gaston, "The Magnitude of Global Insect Species Richness," *Conservation Biology* 5, no. 3 (1991): 283–96.

10. MORE QUESTIONS THAN ANSWERS

2. Pedro Cardoso and Simon R. Leather, "Predicting a Global Insect Apocalypse," *Insect Conservation and Diversity* 12, no. 4 (2019): 263–67; and Dave Goulson, "The Insect Apocalypse, and Why It Matters," *Current Biology* 29, no. 19 (2019): R967–71.
3. Brooke Jarvis, "The Insect Apocalypse Is Here," *New York Times Magazine*, November 27, 2018.
4. Ellen A. R. Welti, Karl A. Roeder, Kirsten M. de Beurs, Anthony Joern, and Michael Kaspari, "Nutrient Dilution and Climate Cycles Underlie Declines in a Dominant Insect Herbivore," *Proceedings of the National Academy of Sciences* 117, no. 13 (2020): 7271–75.

11. THE TEN-TON *T. REX* IN THE HALL CLOSET

1. Lewis H. Ziska and James A. Bunce, "Predicting the Impact of Changing CO_2 on Crop Yields: Some Thoughts on Food," *New Phytologist* 175, no. 4 (2007): 607–18.
2. David Wallace-Wells, *The Uninhabitable Earth: Life After Warming* (New York: Tim Duggan, 2020).
3. Naomi Klein, *The Shock Doctrine: The Rise of Disaster Capitalism* (New York: Picador, 2007).
4. Henning Rodhe, "A Comparison of the Contribution of Various Gases to the Greenhouse Effect," *Science* 248, no. 4960 (1990): 1217–19.

12. WAIT, WHAT?

1. S. P. Long, "Modification of the Response of Photosynthetic Productivity to Rising Temperature by Atmospheric CO_2 Concentrations: Has Its Importance Been Underestimated?," *Plant, Cell & Environment* 14, no. 8 (1991): 729–39.

2. It doesn't mean that plants, especially trees, won't make pollen or other aeroallergens but that the pollen might be sterile.

3. W. J. Arp, "Effects of Source-Sink Relations on Photosynthetic Acclimation to Elevated CO_2," *Plant, Cell & Environment* 14, no. 8 (1991): 869–75.
4. J. T. Baker, L. H. Allen Jr., and K. J. Boote, "Temperature Effects on Rice at Elevated CO_2 Concentration," *Journal of Experimental Botany* 43, no. 7 (1992): 959–64; and Wolfram Schlenker and Michael J. Roberts,

"Nonlinear Temperature Effects Indicate Severe Damages to US Crop Yields Under Climate Change," *Proceedings of the National Academy of Sciences* 106, no. 37 (2009): 15594–98.

5. Tsutomu Matsui, Ofelia S. Namuco, Lewis H. Ziska, and Takeshi Horie, "Effects of High Temperature and CO_2 Concentration on Spikelet Sterility in Indica Rice," *Field Crops Research* 51, no. 3 (1997): 213–19.
6. Xiaoxin Wang, Dabang Jiang, and Xianmei Lang, "Future Extreme Climate Changes Linked to Global Warming Intensity," *Science Bulletin* 62, no. 24 (2017): 1673–80.
7. F. E. Ahmed, A. E. Hall, and M. A. Madore, "Interactive Effects of High Temperature and Elevated Carbon Dioxide Concentration on Cowpea [*Vigna unguiculata* (L.) Walp.]," *Plant, Cell & Environment* 16, no. 7 (1993): 835–42; P. V. Vara Prasad, L. H. Allen Jr., and K. J. Boote, "Crop Responses to Elevated Carbon Dioxide and Interaction with Temperature: Grain Legumes," *Journal of Crop Improvement* 12, nos. 1–2 (2005): 113–55; R. A. C. Mitchell, V. J. Mitchell, S. P. Driscoll, J. Franklin, and D. W. Lawlor, "Effects of Increased CO_2 Concentration and Temperature on Growth and Yield of Winter Wheat at Two Levels of Nitrogen Application," *Plant, Cell & Environment* 16, no. 5 (1993): 521–29; T. R. Wheeler, T. D. Hong, R. H. Ellis, G. R. Batts, J. I. L. Morison, and P. Hadley, "The Duration and Rate of Grain Growth, and Harvest Index, of Wheat (*Triticum aestivum* L.) in Response to Temperature and CO_2," *Journal of Experimental Botany* 47, no. 5 (1996): 623–30; P. V. Vara Prasad, Kenneth J. Boote, and L. Hartwell Allen Jr., "Adverse High Temperature Effects on Pollen Viability, Seed-Set, Seed Yield and Harvest Index of Grain-Sorghum [*Sorghum bicolor* (L.) Moench] Are More Severe at Elevated Carbon Dioxide Due to Higher Tissue Temperatures," *Agricultural and Forest Meteorology* 139, nos. 3–4 (2006): 237–51; Tolentino B. Moya, Lewis H. Ziska, Ofelia S. Namuco, and Dave Olszyk, "Growth Dynamics and Genotypic Variation in Tropical, Field-Grown Paddy Rice (*Oryza sativa* L.) in Response to Increasing Carbon Dioxide and Temperature," *Global Change Biology* 4, no. 6 (1998): 645–56; Ursula M. Ruiz-Vera, Matthew H. Siebers, David W. Drag, Donald R. Ort, and Carl J. Bernacchi, "Canopy Warming Caused Photosynthetic Acclimation and Reduced Seed Yield in Maize Grown at Ambient and Elevated [CO_2]," *Global Change Biology* 21, no. 11 (2015): 4237–49.

13. CRACKS IN THE SYSTEM

1. "UN Report: Nature's Dangerous Decline 'Unprecedented'; Species Extinction Rates 'Accelerating,'" United Nations, May 6, 2019, https://www.un.org/sustainabledevelopment/blog/2019/05/nature-decline-unprecedented-report/.
2. Judith Curry, "Hearing on the Biodiversity Report," *Climate Etc.* (blog), May 22, 2019, https://judithcurry.com/2019/05/22/hearing-on-the-un-biodiversity-report/.
3. Eliza Relman, "Trump's War on Science Is Being Led by a Climate Change Denier Who Compared Carbon Dioxide Pollution to 'Poor Jews Under Hitler,'" *Business Insider*, May 28, 2019, https://www.businessinsider.com/trump-war-on-science-is-led-by-climate-change-denier-2019-5.
4. "Plants Cannot Live on CO_2 Alone," Skeptical Science, last updated July 8, 2015, https://skepticalscience.com/co2-plant-food.htm.

14. SCIENCE SAYS

1. Elizabeth Williamson, "John Boehner: From Speaker of the House to Cannabis Pitchman," *New York Times*, June 3, 2019, https://www.nytimes.com/2019/06/03/us/politics/john-boehner-marijuana-cannabis.html.
2. Chelangat Faith, "The Breadbaskets of the World," *WorldAtlas*, August 1, 2017, https://www.worldatlas.com/articles/the-breadbaskets-of-the-world.html.
3. U.S. Department of Agriculture, *FY 2019 Budget Summary* (Washington, DC: U.S. Department of Agriculture, 2019), https://www.usda.gov/sites/default/files/documents/usda-fy19-budget-summary.pdf.

15. CO_2 IS PLANT FOOD: THE LAST BIT

1. Susan Jaffe, "Republicans' Bills Target Science at US Environment Agency," *Lancet* 385, no. 9974 (2015): 1167–68.
2. Keith O. Fuglie and Paul W. Heisey, *Economic Returns to Public Agricultural Research*, Economic Brief no. 10 (Washington, DC: U.S. Department of Agriculture, Economic Research Service, September 2007).
3. Paul Heisey and Keith Fuglie, "Agricultural Research in High-Income Countries Faces New Challenges as Public Funding Stalls," *Amber*

Waves, May 29, 2018, https://www.ers.usda.gov/amber-waves/2018/may/agricultural-research-in-high-income-countries-faces-new-challenges-as-public-funding-stalls/; and Yu Jin and Wallace E. Huffman, "Measuring Public Agricultural Research and Extension and Estimating Their Impacts on Agricultural Productivity: New Insights from US Evidence," *Agricultural Economics* 47, no. 1 (2016): 15–31.

4. Howard Frumkin and Richard J. Jackson, "We Need a National Institute of Climate Change and Health," *Scientific American*, November 22, 2020.

16. A PERSONAL NOTE

1. Nate Allen, "Nancy Pelosi and Newt Gingrich Commercial on Climate Change," YouTube video, 0:30, uploaded April 18, 2008, https://www.youtube.com/watch?v=qi6n_-wB154.
2. Michael O'Brien, "Gingrich Regrets 2008 Climate Ad with Pelosi," *The Hill*, July 26, 2011, https://thehill.com/blogs/blog-briefing-room/news/173463-gingrich-says-he-regrets-2008-climate-ad-with-pelosi.
3. Tim Lau, "Citizens United Explained," Brennan Center for Justice, December 12, 2019, https://www.brennancenter.org/our-work/research-reports/citizens-united-explained.
4. Isabel Giovannetti, "Inextricably Linked: How Citizens United Halted Climate Action," Common Cause, July 1, 2019, https://www.commoncause.org/democracy-wire/inextricably-linked-how-citizens-united-halted-climate-action/.
5. "Exxon Net Worth," Celebrity Net Worth, accessed October 27, 2021, https://www.celebritynetworth.com/richest-businessmen/companies/exxon-net-worth/#:~:text=Exxon's%20net%20worth%20is%20%24350,production%20based%20in%20Irving%2C%20Texas.
6. Wikipedia, s.v. "Paris Agreement," last modified January 1, 2022, https://en.wikipedia.org/wiki/Paris_Agreement.
7. Alexa Lardieri, "Trump Resort in Ireland Will Build Seawalls to Protect Against Climate Change," *US News*, December 22, 2017, https://www.usnews.com/news/world/articles/2017-12-22/trump-resort-in-ireland-will-build-seawalls-to-protect-against-climate-change.
8. Brady Dennis, "Trump Makes It Official: U.S. Will Withdraw from the Paris Climate Accord," *Washington Post*, November 4, 2019, https://www.washingtonpost.com/climate-environment/2019/11/04/trump-makes-it-official-us-will-withdraw-paris-climate-accord/.

9. "Timeline of Trump's Coronavirus Responses," *Lloyd Doggett, U.S. Representative* (blog), March 2, 2022, https://doggett.house.gov/media/blog-post/timeline-trumps-coronavirus-responses.
10. "Attacks on Science by the Trump Administration," Union of Concerned Scientists, January 20, 2017, https://www.ucsusa.org/resources/attacks-on-science.
11. Ian Kullgren, "How Perdue's Power Benefits His Friends," *Politico*, March 13, 2017, https://www.politico.com/story/2017/03/donald-trump-sonny-perdue-agriculture-235982.
12. Charles P. Pierce, "Everything Is So Insanely Dumb and It's Going to Kill Us," *Esquire*, June 25, 2019, https://www.esquire.com/news-politics/politics/a28187258/agriculture-secretary-sonny-purdue-climate-change/.
13. Ben Guarino, "Trump Administration Plans to Move USDA Research Divisions Despite Concerns," *Washington Post*, April 25, 2019, https://www.washingtonpost.com/science/2019/04/25/trump-administration-plans-move-usda-research-divisions-despite-concerns/.
14. Eric Katz, "Mulvaney: Relocating Offices Is a 'Wonderful Way' to Shed Federal Employees," *Government Executive*, August 5, 2019, https://www.govexec.com/workforce/2019/08/mulvaney-relocating-offices-wonderful-way-shed-federal-employees/158932/.
15. Jesse Naranjo, "200 ERS Vacancies After Kansas City Move," *Politico*, February 27, 2020, https://www.politico.com/news/2020/02/27/200-ers-vacancies-after-kansas-city-move-117764.
16. Keith Fuglie, Matthew Clancy, Paul Heisey, and James MacDonald, "Research, Productivity, and Output Growth in U.S. Agriculture," *Journal of Agricultural and Applied Economics* 49, no. 4 (2017): 514–54.
17. "Pacific West Area Map," Agricultural Research Service, U.S. Department of Agriculture, accessed October 26, 2021, https://www.ars.usda.gov/pacific-west-area/docs/map/.
18. Liz Crampton, "Sam Clovis Is Leaving," *Politico*, May 3, 2018, 2021, https://www.politico.com/story/2018/05/03/sam-clovis-leaving-usda-518109.
19. Scott H. Hutchins, "Natural Products for Crop Protection: Evolution or Intelligent Design," in *Discovery and Synthesis of Crop Protection Products*, ed. Peter Maienfisch and Thomas M. Stevenson (Washington, DC: American Chemical Society, 2015), 55–62.
20. Chunwu Zhu, Kazuhiko Kobayashi, Irakli Loladze, Jianguo Zhu, Qian Jiang, Xi Xu, Gang Liu, et al., "Carbon Dioxide (CO_2) Levels This Century Will Alter the Protein, Micronutrients, and Vitamin Content

of Rice Grains with Potential Health Consequences for the Poorest Rice-Dependent Countries," *Science Advances* 4, no. 5 (2018): eaaq1012.

21. Helena Bottemiller Evich, "'It Feels Like Something Out of a Bad Sci-Fi Movie,'" *Politico*, August 5, 2019, https://www.politico.com/story/2019/08/05/ziska-usda-climate-agriculture-trump-1445271.
22. Merrit Kennedy, "As Carbon Dioxide Levels Rise, Major Crops Are Losing Nutrients," NPR, June 19, 2018, https://www.npr.org/sections/thesalt/2018/06/19/616098095/as-carbon-dioxide-levels-rise-major-crops-are-losing-nutrients; and Brad Plumer, "How More Carbon Dioxide Can Make Food Less Nutritious," *New York Times*, May 23, 2018, https://www.nytimes.com/2018/05/23/climate/rice-global-warming.html.
23. "Request for Information (RFI): Climate Change and Health," NOT-ES-21-009, U.S. National Institutes of Health Grants and Funding, July 30, 2021, https://grants.nih.gov/grants/guide/notice-files/NOT-ES-21-009.html (first grant announcement from the NIH on climate change and public health).
24. Naomi Oreskes and Erik M. Conway, "Defeating the Merchants of Doubt," *Nature* 465, no. 7299 (2010): 686–87.

INDEX

INDEX

INDEX